R O International Real Option Workshop
06. - 08.05.2002 Turku, Finland

Proceedings of the
International Real Option Workshop
06.-08.05.2002 Turku, Finland

Edited by Mikael Collan

RO International Real Option Workshop
W 06. – 08.05.2002 Turku, Finland

DAY 1

Chairman: Christer Carlsson

REGISTRATION AND OPENING   09.00 – 11.00

Registration and coffee
Opening
The WAENO Research program

SESSION 1   11.00 – 12.00
Keynote Speaker
Lenos Trigeorgis

Lunch   12.00 – 13.00

SESSION 2   13.30 – 15.00
Janne Gustafsson and Ahti Salo
Philipp Baecker
M-Real Case

Afternoon coffee   15.00 – 15.30

SESSION 3   15.30 – 17.00
Arvind Girotra
Markku Kallio
Outokumpu Case

EVENING PROGRAM   19.30 –

DAY 2

Chairman: Juha Paappanen

SESSION 1   09.00 – 10.30
Marie–Laure Guillerminet
Marcus Dimpfel and René Algesheimer
Rautaruukki Case

Morning coffee   10.30 – 10.45

SESSION 2   11.00 – 12.00
Keynote Speaker
Alexander Bukhvalov

Lunch   12.00 – 13.00

SESSION 3   13.00 – 15.30
Christer Carlsson and Robert Fullér
Mikael Collan
Péter Majlender, Markku Heikkilä

Afternoon coffee   15.30 – 15.45

SESSION 4   15.45 – 17.15
Keynote Speaker
Marco Guimaráes Dias
Francisco Alcaraz
Fortum Case

EVENING PROGRAM   19.30 –

DAY 3

Chairman: Robert Fullér

SESSION 1   09.00 – 10.30
Keynote Speaker
Luis Alvarez
Anett Mehler–Bicher

Morning coffee   10.30 – 10.45

SESSION 2   10.45 – 12.15
Matts Rosenberg
Christer Carlsson
Svante Olofsson

Lunch   12.15 – 13.15

SESSION 3   13.15 – 15.00
Mikael Collan
Shuhua Liu
EUNITE IBA E and Discussion of Future
Research Directions

CLOSING OF THE WORKSHOP 15.00

eunite

TUCS

R□ International Real Option Workshop
W 06. - 08.05.2002 Turku, Finland

ORGANISING COMMITTEE

Chairman	Professor Christer Carlsson, Åbo Akademi University
Deputy Chairman	Professor Robert Fullér, Eötvös Lórand University
Secretary	Sirpa Nummila, IAMSR, Åbo Akademi University
Members	Mikael Collan, IAMSR, Åbo Akademi University
	Markku Heikkilä, IAMSR, Åbo Akademi University
	Piia Hirkman, IAMSR, Åbo Akademi University
	Péter Majlender, IAMSR, Åbo Akademi University

CORPORATE PARTNERS

m·real

Fortum

RAUTARUUKKI

DAY 1

Monday 06.05.

Chairman   Christer Carlsson

REGISTRATION AND OPENING

09.00 – 10.15 Registration and coffee

10.15 – 10.30 Opening
Professor Christer Carlsson, IAMSR, Åbo Akademi University

10.30 – 11.00 The WAENO Research program

SESSION 1

11.00 – 12.00 Keynote Speaker
Lenos Trigeorgis, Real Options Group

12.00 – 13.00 Lunch

SESSION 2

13.30 – 14.00 Janne Gustafsson and Ahti Salo
"Applying Real Options to the Valuation of Industrial R&D Projects"

14.00 – 14.30 Philipp Baecker
"Real Options Valuation for Drug R&D"

14.30 – 15.00 M–Real Case

15.00 – 15.30 Afternoon coffee

SESSION 3

15.30 – 16.00 Arvind Girotra
"Valuing Flexibility"

16.00 – 16.30 Markku Kallio
"Real Option Valuation via Stochastic Optimization"

16.30 – 17.00 Outokumpu Case

19.30 – *Evening Program* – Dinner Cruise into the Archipelago
onboard the SS Ukkopekka

DAY 1

SESSION 2

Janne Gustafsson and Ahti Salo
"Applying Real Options to the Valuation of Industrial
 R&D Projects"

Philipp Baecker
"Real Options Valuation for Drug R&D"

*M–Real Case   m·real

*No Material

Applying Real Options to the Valuation of Industrial R&D Projects – A Comparative Case Study

Janne Gustafsson and Ahti Salo
Systems Analysis Laboratory
Helsinki University of Technology
P.O. Box 1100, 02015 HUT, Finland

Abstract: We report experiences on the applicability of real options to the valuation of industrial R&D projects, based on a multi-client project in which real options and three other approaches were explored in six case studies at major enterprises in Finland. In these case studies, uncertainties and technological discontinuities related to the projects made it impractical to use continuous stochastic processes for modelling purposes. In contrast, traditional approaches (i.e., decision trees, scenario-based net present value) were found helpful both in problem structuring and in illustrating uncertainties and associated options. We conjecture that these approaches may be better suited to the valuation of industrial R&D projects than real options modelling, especially if the value of the project's results cannot be linked to market information.

Keywords: Real options, decision analysis, financial modelling.

DAY 1

SESSION 3

Arvind Girotra
"Valuing Flexibility"

Markku Kallio
"Real Option Valuation via Stochastic Optimization"

*Outokumpu Case

*No Material

VALUING

By

Arvind Girotra

B.E.(IIT-Roorkee), MBA(IIT-Delhi)

Functional Consultant ICICI

Infotech Services Limited

2nd Floor, Zenith House

Keshavrao Khadye Marg, Mahalaxmi

Mumbai- 400 034. India

E-mail: arvindgirotra@rediffmail.com

Phone: 91-22-490 6607

Abstract

Today's business environment is rife with challenges – ranging from global competition to fast changing technology to volatility in the financial markets – that have major impact on the survival of an organization. Driven by uncertainty and chaos in business environment, now the clear imperative is greater focus on enhancing the flexibility by building in options in real situations.

Traditional decision making approaches like NPV and Decision Trees have limitation as far as valuing flexibility is concerned. However, thanks to pioneer work by Black, Scholes, option pricing theory has come to rescue of finance managers. Real option valuation is important in situations of high uncertainty where management can respond flexibly to the new information and where project value without flexibility is near breakeven. Real option valuation not only captures all situations but also the flexibility the management has in reacting to such situations.

By ignoring the option value, management may ignore genuine investment opportunities just because they were undervalued. In the world of cut-throat competition, this thought is frightening and is a threat to survival.

The paper attempts to highlight some areas where flexibility is at work and methods to value flexibility with case studies to make the concepts more comprehensible.

Keywords: flexibility, options, valuation, models, net present value, decision tree

---

1.0 Introduction

Today's business environment is rife with challenges – ranging from global competition to fast changing technology to volatility in the financial markets – that have major impact on the survival of an organization.

In the business environment today, following forces are at work:

- *Protectionism is out and competition is in*

1990s have shown Indian Business an exit door to protectionist policy of the government. There has been a marked shift in policy and the tilt is clearly towards competition. Global companies and their products have entered the Indian markets and the playing field is much more level than before. The competition is not only in prices but also in technology and the margins are on a decline.

- *Information Technology revolution has set in*

Information Technology has accelerated the pace of change. Information is much more widely available and at an ever decreasing cost. Competitive advantage now lies not in assembling of information, but rather in using it for decisions.

- *Volatility rules*

The financial markets worldwide have been very volatile; India is no exception. The exchange rates and interest rates have been volatile. Many new financial products have been and are being launched.

All these forces have spelt two things for sure for corporates:

✓ One, uncertainty is very high, changes are fast and there is a need for a flexible enterprise. There is a need to be flexible, agile and responsive in all functions of a corporate. The decisions have to be such that they leave managers with more options to take care of uncertain future.

✓ Second, the shareholder value is at risk. There is a need to manage value and develop a value based management framework. Management should assess the shareholder value creation in whatever the company does. Flexible approach may help in the process.

If we see both the imperatives listed above in conjunction, then it throws up an interesting question, "How much is flexibility worth?" Or in other words, "How much does it cost to retain flexibility?"

The paper attempts to highlight some areas where flexibility is at work and methods to value flexibility with case studies to make the concepts more comprehensible.

2.0 Flexibility at Work: The Real Options

Two strategies can be used for dealing with uncertainty: (1) Anticipation (2) Resilience [1]. If risks can be anticipated because they are predictable, then the most effective and least costly approach is often to construct a specialized and relatively inflexible system that works best in the anticipated environment. Alternatively, if risks cannot be anticipated, a resilient system with a great deal of flexibility becomes the best approach. In a resilient system, flexibility in a corporate can be classified as follows:

✓ *Asset side flexibility*

✓ *Liability side flexibility*

Asset side flexibility comprises of different type of options available on assets (also known as real options). These can be of following types:

➤ *Growth options* : These include options of scaling up, switching up or scoping up a project. For example, management may choose to build production capacity in excess of expected level of output so that it can produce at a higher rate if the product is more successful than was originally anticipated. Because expansion gives management the right, but not the obligation, to make additional follow on investment, a project that can be expanded is much more worth than same project without the flexibility to expand.

➤ *Deferral/ Learning Options.* These include options to delay investment until more information or skill is acquired. Such options could be natural resource development (coal mining/ oil exploration etc.) and pharmaceutical R&D. For example, the owner of exploration rights of an undeveloped oil reserve can defer the development process until the oil prices rise. In other words, the managerial option implicit in holding an undeveloped reserve is in fact a deferral option. Because, the deferral investment option gives management a right, but not obligation, to make investment to develop the property, a project that can be

----

deferred is worth more than the same project without the flexibility to defer development.

➢ *Disinvestment/ Shrink/ Abandonment Options.* These include options to scale down/ switch down or scope down a project. For example, option to abandon an open pit coal mine sets a lower bound on the value of project. A project that can be liquidated is worth more than the same project without the possibility of abandonment.

Liability side flexibility vests in the form of options on account of convertible debt/ warrants/ debentures with call options. These directly impact the cost of capital of the company. For example, a financial institution borrows long term and lends short term. In a falling interest rate scenario, such company would be locked in with higher cost of funds and would have deployed such funds at a lower rate leading to losses. In such scenario, imagine, if this company had included a call option in instruments issued at the time of borrowing. In doing so, though it would have incurred a little extra cost but would have saved itself of the current imbroglio.

The real options discussed above take care of uncertainty surrounding technology, market acceptance, prices etc. Options revolving around multiple sources of uncertainty are called rainbow options [2].

The real options could be one at a time or more than one at a time (referred to as compound options [2] such as pharmaceutical R&D projects will have an option to commercialize the product and another option to engage in subsequent R&D work to develop future generation of related products.).

Such options create value, that is, cost of option is lesser than the benefit it provides.

3.0 Traditional Decision Making Approaches

Managers use variety of techniques to evaluate investments under uncertainty. Some of these techniques are: (1) Net present value (NPV) (2) Internal rate of return (IRR) (3) Earnings growth (4) Decision Tree Analysis (DTA) (5) Economic Profit.

Out of above, Discounted cash flow (DCF) techniques have been more popular viz. NPV, IRR and DTA. Earnings growth and Economic Profit being accounting based measures are the least used.

Within DCF techniques, NPV has certain superiority over IRR on account of various reasons [7]. This leaves us with NPV and DTA practically.

3.1 Net Present Value and Flexibility

In traditional NPV analysis, future cashflows are forecasted, discounted at a risk-adjusted rate and current investment cost is subtracted to estimate net present value of the project. Projects with positive NPV are said to create value and are accepted; negative NPV projects are not accepted.

Let us take an example, Pharmaceutical R&D projects are carried out in stages and further investments are done considering the feasibility of the project at the future date. In case state of nature is bad, the project is abandoned. In traditional NPV analysis, such options at a later date to abandon or defer are not considered. Hence it overlooks management's flexibility to alter the course of project in response to changing market conditions. In other words, traditional NPV analysis assumes that companies hold investments passively. However, decisions on investments are dynamic (vary with time) and a function of various factors such as market conditions.

3.2 Decision Tree Analysis and Flexibility

In another example, say, project requiring investment of Rs 65 mn has payoff of Rs 120 mn in good state of nature and Rs 30 mn in bad state of nature respectively with equal likelihood of happening.

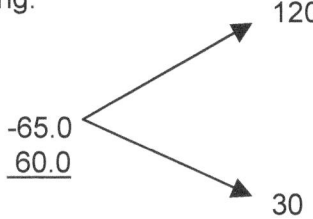

Say, project has opportunity cost of capital as 25%(found by a comparable security with similar payoff over one year-diagrammed below). This cost takes care of the risks embedded in the payoff.

Price, t=0   Price, t=1

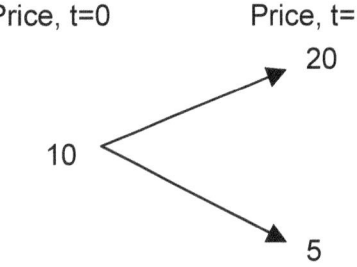

The NPV of the project is calculated as follows: NPV = -65 + 0.5 {(120 + 30)/1.25}

> = -65 + 75/1.25

> = -5.0

Using DTA, the NPV of the project is Rs –5.0 mn and hence not acceptable.

Now, let us complicate the situation. Assume the company has option to wait for one year (one-year risk-free rate as 6%) and invest only if good state of nature takes place. In alternative case, the license expires. This will dramatically alter the payoffs. The point to note is that investment is required only if the good state of nature takes place, in other words, the cash outflow takes place in period 1. The future value of cash outflow would be Rs 65.0 *(1.06) mn. We can calculate the value created in period 1 which will be either 0 (bad state of nature and no investment will be made) or 120 – 65*(1.06)=Rs 51.1 mn (good state of nature)

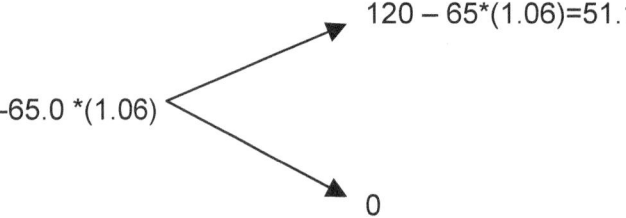

Now, we can't apply the same discount rate as applied earlier as the payoffs have completely changed or in other words, the risk character has completely changed.

If same rate as earlier is applied, the NPV using DTA would be:

> = (0.5*51.1 + 0.5*0)/1.25

> = Rs 20.44 mn

This shows that DTA does capture flexibility but is unable to establish a proper risk-adjusted discount rate. Next section on option pricing methods explains how option-pricing methods include both flexibility as well as a proper risk adjusted discount rate.

15

4.0 Option Pricing Theory

4.1 What is an option?

An option provides the holder with the right to buy or sell a specified quantity of an underlying asset at a fixed price (strike price) on or before the expiration date of the option. Since it is a right and not an obligation, the holder can choose not to exercise the right and allow the option to expire. There are two types of options: call options and put options.

Call options

A call option gives the buyer of an option the right to buy the underlying asset at a fixed price at any time prior to or on the expiration date of the option. The buyer pays a price (also known as premium) for this right.

Put options

A put option gives the buyer of an option the right to sell the underlying asset at a fixed price at any time prior to or on the expiration date of the option. The buyer pays a price for this right.

4.2 Determinants of Option Value

Following table clearly brings out the effect of various variables that determine the value of an option.

S.No.	Variables	Value of Call option	Value of Put option
1.	Current value of underlying asset (+)	(+)	(-)
2.	Volatility in the value of underlying asset (+)	(+)	(+)
3.	Dividends paid on the underlying asset(+)	(-)	(+)
4.	Strike price of option(+)	(-)	(+)
5.	Time to expiration of option(+)	(+)	(+)
6.	Risk-less interest rate corresponding to the life of option(+)	(+)	(-)

4.3 Option Pricing Models

Option pricing theory has made vast strides since 1972, when Black and Scholes [8] published their path breaking paper providing a model for valuing European options.

Applications have proliferated, particularly in securities market, where theory held up remarkably well when tested against actual prices.

The idea at work in option pricing is that risk-free arbitrage is not possible. Put simply, no arbitrage means that securities with exactly same risk return profiles should be identically priced. If you can describe the payoffs on one risky security and then a build a portfolio of other securities with exactly the same payoff, the price of both must be the same. If the prices were not identical, arbitrage could be possible. Here we shall restrict ourselves to binomial and black-scholes model only.

4.3.1 Binomial One Step Model
Law of one price

The price can move to two possible prices. Let us consider the example in section 3.2 equivalent to a call option. Assume a security, which has identical payoff as the asset underlying real option. The security is diagrammed below:

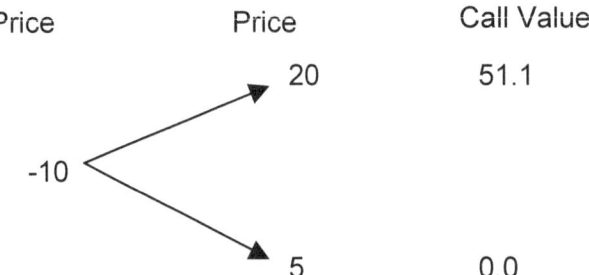

Price	Price	Call Value
	20	51.1
-10		
	5	0.0

Creating a replicating portfolio by combination of the security and risk-free lending/ borrowing to create same cashflows as option being valued.

Let

S – Current price of security,

Δ - no. of units of security,

B - risk-free borrowing,

r – risk-free rate of return,

t – time period of option.

Matching payoffs in two states of nature:

$20\,\Delta -- B\,e^{rt} = 51.1$ ..... (1)

$5\,\Delta -- B\,e^{rt} = 0.0$ ..... (2)

Solving (1) and (2) we get

B= 16.0414

$\Delta = 3.4067$

The cost of portfolio today is

$= S\Delta - B$

$= 10 * 3.4067 - 16.0414$

$= 18.03$

The value of the option should be equal to cost of setting up of portfolio otherwise arbitrage will apply. Hence the cost of license should be Rs 18.03 mn. Now, compare this with the value obtained from DTA. Clearly, DTA overvalued the investment because it could not correctly apply the risk-adjusted discount rate.

In a multi-period binomial process, the valuation has to proceed iteratively starting with the last time period and moving backwards in time until current point in time. The portfolio replicating the options are created at every step and valued, providing the value for the option in that time period. The final output from the binomial option-pricing model is that value of the options that can be stated in terms of replicating portfolio composed of Δ units of underlying asset and risk-free borrowing / lending.

Setting up a risk-less portfolio

Consider a portfolio consisting of a long position in Δ units of security and a short position in one call option (award of license).

Δ units are such that they along with one option make the portfolio riskless whichever way the price of security moves, the portfolio value remains the same.

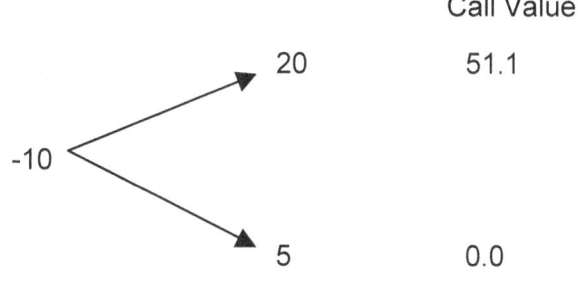

Call Value

20   51.1

-10

5   0.0

That is,

Value of portfolio in good state of nature = Value of portfolio in bad state of nature

$20\,\Delta - 51.1 = 5\,\Delta - 0$

or $\Delta = 3.4067$

A riskless portfolio is therefore,

Long 3.4067 units and short one option.

Value of portfolio in favorable state of nature:

$=20 * 3.4067 - 51.1 = 17.03$

Value of portfolio in unfavorable state of nature:

$=5*3.4067 = 17.03$

Regardless of the state of nature, value of the portfolio is always Rs 17.03 mn at the end of life of option.

Riskless portfolio must, in absence of arbitrage opportunities, earn risk-free rate of interest. It follows that value of portfolio today must be the present value of 17.03 or $17.03 \, e^{-.06}$ (where r =0.06, t=1)

The current value of security is 10. Suppose that option price is 'f'. The value of portfolio, today, is therefore,

$3.4067*10 - f = 17.03 \, e^{-.06}$

$f=18.03$

This shows that in absence of arbitrage opportunities, the current value of the option must be Rs 18.03 mn. If the value of option were more than the portfolio would cost less than $17.03 \, e^{-.06}$ to set up and would earn more than the risk-free rate. If the value of option were less than Rs 18.03 mn shorting the portfolio would provide a way of borrowing money at less than risk-free rate.

Risk Neutral Valuation

This principle states that we can with complete impunity assume that the world is risk neutral when pricing options and other derivatives. In such a world, investors require no compensation for risk and expected return on all securities is risk free interest rate.

Assuming, p, as probability of occurrence of favorable state of nature. As per this, $20p + 5(1-p) = 10 \, e^{0.06}$

$p=0.3746$

At the end of one period, the call option has p probability of being worth 51.1 and 1-p probability of being worth 0. Its expected value is therefore,

$=0.3746*51.1$

$=19.14$

Discounting at risk free rate of interest, value of option today is,

$= 19.14 \, e^{-0.06}$

$= 18.03$

This is same as obtained earlier with no arbitrage argument. Hence, prices we get are correct not just in a risk neutral world but in other worlds as well [3].

4.3.2 Black-Scholes Model

The binomial model is a discrete time model for asset price movements, with a time interval (t) between price movements. As the time interval is shortened, the limiting distribution, as t approaches 0, can take one of two forms.

(a) If price changes become smaller, as t approaches 0, the limiting distribution is the normal distribution and price process is a continuous one.

(b) If price change remains large, as t approaches 0, the limiting distribution is the Poisson distribution, that is, a distribution which allows for price jumps.

The black-scholes model applies when the limiting distribution is the normal distribution.

Assumptions of the Model (for non-dividend paying asset)

➢ The log of return on underlying asset is normally distributed over time.

➢ The price of underlying asset follows the weiner process. The weiner process is akin to the brownian motion in physics. The change in variable is normally distributed and function of drift rate and variance rate.(for detailed explanation see Chapter 10 of 'Futures, Options and other derivatives' by J.C.Hull).

➢ The short selling of underlying asset with full use of proceeds is permitted.

➢ There are no transaction costs or taxes.

➢ There are no dividends during the life of option.

➢ There are no risk-less arbitrage opportunities.

➢ Asset trading is continuous.

➢ The risk-free rate of interest is constant.

The Model

The value of a call option in the Black-scholes model can be written as a function of following variables:

S= Current value of underlying asset

K= Strike price of option

t= Time to expiration of the option.

r = Risk-less interest rate corresponding to the life of option.

σ^2 = Variance in the ln(returns) on the underlying asset.

The model itself can be written as : Value of the call, $C = S N(d_1) - K e^{-rt} N(d_2)$ $d_1 =$ $(\ln(S/K) + (r + \sigma^2/2)^*t)/(\sigma^*t^{1/2})$

$d_2 = d_1 - \sigma^*t^{1/2}$

Long term options (with constant yield, y)

$C = S e^{-yt} N(d_1) - K e^{-rt} N(d_2)$

$d_1 = (\ln(S/K) + (r - y + \sigma^2/2)^*t)/(\sigma^*t^{1/2})$

$d_2 = d_1 - \sigma^*t^{1/2}$

5.0 Application of Black-Scholes Model to Value Real Options

5.1 Case Study: Valuing Natural Resource Investments as Options
The Framework

In a natural resource investment, the underlying asset is the resource and value of the asset is based upon two variables: quantity of resource that is available in the investment and the price of the resource. In such investments, there is a cost associated with developing the resource and the difference between the value extracted and the cost of development is the profit to owner of the resource.

Say, cost of development is K, and the estimated value of the resource is S, then payoff on natural resource investment will be:

= S – K   if S>K

= 0   if S<= K

Thus, the investment in a natural resource option has a payoff similar to that of a call option.

The Case

Company ABC Limited has been in the business of oil exploration for last few decades. ABC wants to bid for the rights to exploit an oil property 'Panna' for next twenty-five years. ABC earlier got a study conducted on the oil field by two well-known teams of geologists and the estimated oil reserves are 100 mn barrels of oil. Taking past experiences of exploration in similar oil fields, the management is expecting the present value of development cost to be around Rs 600 per barrel. Also the development lag is estimated to be one year. Once developed, the net production revenue each year will be 4% of the value of the reserves.

After analyzing the past trends and forecasts, the operations and marketing team expects marginal present value (price-marginal cost) per barrel of oil to be around Rs 600 as well. The risk less rate is 10%.

Current value of asset: marginal present value X no of barrels/ ((1 + dividend yield)^(lag period for development))

=Rs 600 x 100 mn/1.04

= Rs 57962.3 mn

(value of the developed reserve discounted back the development lag period at dividend yield, 4%)

Present Value of development cost: 600 x 100 =Rs 60000 mn

This meant a clear no to going ahead on exploration of oil field.

However, the top management was of the opinion that variability in the oil prices (whenever the price of oil rises, the company could go ahead with the exploration and make profits) made a case for some value for the rights. What could be the value?

Keeping variability of available reserves of oil and variability in oil prices as key factors, the variance in ln(oil price) was estimated at 0.05. Then, Black-Scholes model was applied to arrive at the value of the rights (value of call):

S = Rs 57962.3 mn

K = Rs 60000 mn

r = 0.10

y =0.04

----

t = 25 years

$\sigma^2 = 0.05$

d1= 1.87   N(d1)= 0.9693

d2= 0.752   N(d2)= 0.7740

Value of the call: Rs 16856.5 mn

We can see that the oil reserve not viable at current prices, still is a valuable property because of its potential to create value if the oil prices go up.

5.2 Case Study: Valuing Product Patents as Options

The Framework

A product patent provides the firm with the right to develop the product and market it. It will do so only if the present value of the expected cash flows from the product sales exceed the cost of development. If this does not occur, the firm can shelve the patent and not incur any further costs.

Say, the present value of costs of developing a product is K, and present value of expected cashflows from development is S, then the payoff from owning a product patent will be:

=S – K   if S>k

=0   if S<=K

The Case

Pure Pharmaceuticals has the patent rights, for next twenty five years, to a product which is a potential cure for AIDS(All Immune Deficiency Syndrome). The initial investment to develop the product is estimated to be around Rs 20 bn.

Considering the most probable scenario, the present value of expected cash inflows is Rs 16 bn. Hence, at current levels, the patent seemed to have no value.

For the following reasons, the management thought that patent could become a valuable project at a future date:

1.   Variance in the expected cash inflows leading to scenarios where prices may move up.

2.   Positive development in the competitive environment.

3.   Better marketing synergies.

4.   Technological changes that bring the cost and development time down.

Keeping above in mind, management decided that patent has value. What could be the value?

With the above factors as key variables, the variance in present value of cash inflows was estimated at 0.05. Risk less rate is 10%.

Then, Black-Scholes model was applied to arrive at the value of the patent (value of call):

S = Rs 16 bn

K = Rs 20

r = 0.10

y =0.04, Percentage of patent value derived realized in one year

t = 25 years

σ^2 = 0.05 d1= 1.701

$\qquad\qquad\qquad\qquad\qquad\qquad$ N(d1)= 0.9555

d2= 0.583 $\qquad\qquad\qquad\qquad\qquad$ N(d2)= 0.72

Value of the patent/ call: Rs 4.44 bn

We can see that the patent not viable at current prices, still is a valuable product when viewed as option because of its potential to create value in favorable state of nature.

6.0 Limitations of Black-Scholes Model

Following are the limitations of the Black-Scholes model when applied to real options [4]:

➢ Real options are not traded and hence arbitrage argument does not hold true. For example, product patents are not actively traded and hence suffer from this limitation.

➢ The price of the real asset may not follow a continuous process. One solution is to use models, which allow for price jumps.

➢ The assumption of a known variance and it being constant over the life of option may not hold in long term real options. Estimating variance is difficult in real options.

➢ Liability side options are not taken care of.

We shall now try to deliberate on how to treat the liability side options, which are the order of the day in today's scenario. With the kind of uncertainty prevailing, the liabilities issued by the company have embedded options. A non-exhaustive list includes equity of leveraged company, warrants, callable or convertible debt and preferred stock, variable rate loans with caps or floors, guaranteed lines of credit, operating leases and executive stock options. These securities provide flexibility in the hands of the company management in coping up with stress scenarios and save the organization from the low probable high loss. This flexibility comes at a cost and directly impacts the cost of capital [1].

Black-Scholes Model, which worked quite well in valuing real options of asset side has limitations when applied to real liability side options. The issues, which limit the applicability of Black-Scholes model in valuing long term interest rate options, are [5]:

1. Assumption that the interest rate is constant leads to bias when valuing long term options on long term bonds.

2. The second problem is that the price of the bond will not be lognormally distributed, because it is forced to par value at the bonds' maturity.

3. If the bond is tied to a price of par at maturity, then volatility of its price must fall as it matures. In Black-Scholes model, volatility is assumed to be constant.

These problems can be overcome by using bonds' forward price and the relevant volatility of that forward price. This in turn would require further assumptions on yield and volatility.

The more sophisticated models work on rate as the state variable rather than the price of bond talked hitherto. If a bond has no risk of default, then its price is uniquely determined by the spot rates of interest. Once these rates are known, any bonds' price can be calculated.

The pattern of spot rates at any point in time is known as term structure of interest rates. Various models have been proposed such as one factor and two factor models. One-factor models assume short-term interest rates as the only source of uncertainty, however, two factor models assume both short term and long term interest rates as source of uncertainty.

Two factor models are quite complex and are not pursued here. Some of the one factor models that use binomial /trinomial tree to model interest rates are Ho-Lee(1986), Heath etal (1988, 1989), Hull-White(1990) etc. Black-Derman-Toy(BDT) which is among the simpler of these models is presented here. Next section details how BDT creates a term structure of interest rates using binomial tree and then derive the value of embedded option.

7.0 The Black-Derman-Toy Model

Features

✓ Short-term interest rates are lognormally distributed. That is, the log of returns on short term rates are normally distributed.

One period Rate Tree (t=1, volatility=σ)

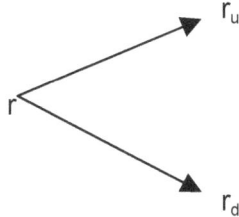

Ln $(r_u/r) = \sigma$ .... (a)

Ln$(r/ r_d) = \sigma$ .... (b)

Combining (a) and (b), we get

$r_u / r_d = e^{2\sigma}$ ....   (1)

This ensures a decreasing rate of depreciation in the lower part of the tree as compared to the upper part and negative interest rates will not occur.

✓ The volatility of interest rate depends upon the term and is different for each term.

✓ The inputs to the model are today's zero coupon yields (spot rates) for various maturities and the respective todays' volatility.

✓ Using equations, the forward rates are calculated. We have already seen the relationship between rates and volatility. Second relationship is derived by allowing the bond price to converge. Say, the market price of 2 year bond is M.

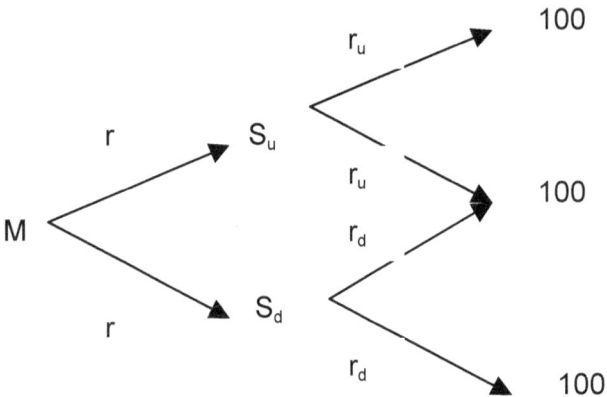

$(0.5* S_u + 0.5* S_d)/(1+r) = M$ ....   (2)

$S_u=100/(1+ r_u)$ ....   (3)

$S_d=100/(1+ r_d)$ ....   (4)

Solving (1), (2), (3) and (4), we get r_u and r_d which are second period forward rates. Taking these as base and moving ahead with same rate tree and price tree constraints, the forward rates for various periods can be calculated.

✓ Once the term structure is evolved, option on the bond can be valued by discounting the option payoff on various nodes of the tree to the present value. We shall take a case study to give a clear step by step option valuation using BDT model.

7.1 Case Study: Valuation of Bond with Embedded Call Option

The zero coupon yields and their respective volatility are as under:

Maturity (years)	Zero Coupon Yields(%)	Volatility(%)
1	10	20
2	11	19
3	12	18
4	12.5	17
5	13	16

Following steps listed in appendix-I and moving further, we can evolve the following term structure of interest rates (bold). Then discounting the prices (starting from period 5 backwards) from two scenarios at forward rates derives the prices for the earlier period. Price in period 4 fo forward rate of .2553 will be computed as (0.5*100 +0.5*100)/1.2553=79.66

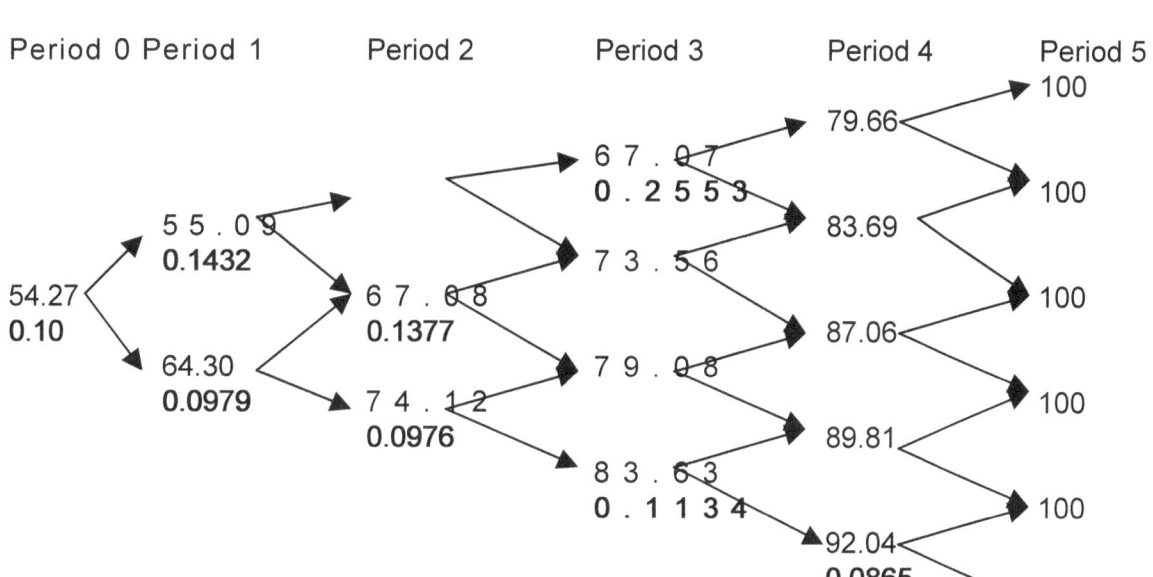

Company XYZ issues a zero coupon bond with a par value of Rs 100. The bond does not have any options. The company will be able to sell the bond in the market at Rs 54.27. The yield will be 13%. Now, let's assume that company keeps an option to call the bonds at the end of third year if the price is greater than Rs75 in which case it would be profitable for the company to buyback the bonds and reissue in the falling rate scenario. This is akin to company buying a call option, which has been written by the investor. In scenario, where price increases over Rs 75, the profits of investor will be limited to a price of Rs 75 as the company will call the bond.

The valuation of the callable bond will be as under:

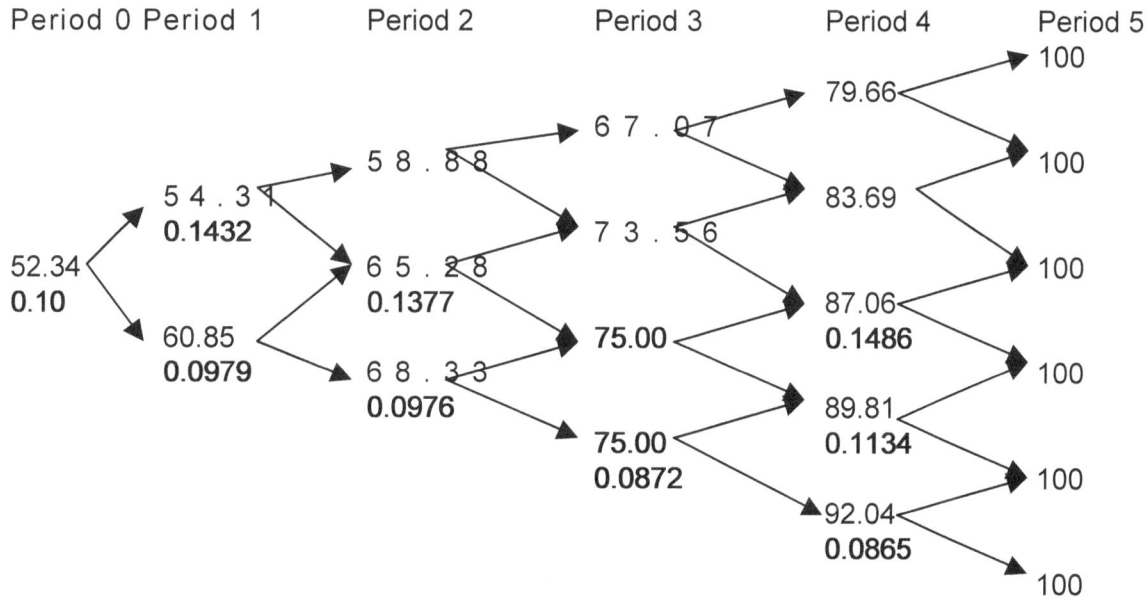

The value of callable bond is Rs 52.34 or Rs 1.93 is the call option premium paid by the company to the investor. The yield works out to 13.8%.

The yield in case of non callable bond was 13% whilst in callable bond it has increased to 13.8%($((100/52.34)^{(1/5)}-1)*100)$. thus, we can see that options on liability side have direct impact on cost of capital of the company.

We can carry out a similar analysis for a bond with discrete coupon payments. In that case, the cash flow in t year will have to be discounted by t-1 forward rate to arrive at market price in t-1 year. Moving backwards we can value the bond at t=0.

When the callable bonds have a higher cost of capital, the question arises as to why such bonds are issued/ As has been explained earlier, such options provide flexibility to the organization in reacting to situations which could not be foreseen. An example in Indian context is the issuance of long term bonds by financial institutions (FIs) in high interest rate scenario. This led to FIs getting locked in with high cost funds however assets in the meantime repriced to lower interest rates as rates fell over time resulting in heavy erosion of the bottomline. The FIs, which issued callable bonds, saved their skin, thanks to the flexibility they had built in.

8.0 Conclusion

Driven by uncertainty and chaos in business environment, now the clear imperative is greater focus on enhancing the flexibility by building in options in real situations. Traditional decision making approaches like NPV and Decision Trees have limitation as far as valuing flexibility is concerned. NPV does not capture the mechanics of flexibility and decision tree methodology gives no guidance on how to choose discount rates or adjust it for risk or leverage.

Key Criteria for decision-making tools [6] Method / Tool

	Cash Flow based	Risk Adjusted	Multi Period	Capture Flexibility
Real Options Value	√	√	√	√
NPV/DCF	√	√	√	X
Decision Tree	√	X	√	√
Economic Profit	X	√	X	X
Earnings Growth	X	X	X	X

Real option valuation is important in situations of high uncertainty where management can respond flexibly to the new information and where project

---

without flexibility is near breakeven. Real option valuation not only captures all situations but also the flexibility the management has in reacting to such situations. We have discussed various models for valuing real options in this paper. The intention is to make the underlying concept vivid. At this time, it is important to note that use of complex models like Black-Scholes and others should be preceded by an understanding of underlying assumptions and their validity in case context.

By ignoring the option value, management may ignore genuine investment opportunities just because they were undervalued. In the world of cut-throat competition, this thought is frightening and is a threat to survival. We have seen that option valuation is important not only to the asset side but also to the liability side of company.

The greater issue is to consider the value of options. We may begin by being little less correct but at least a beginning should be made.

References

1. Copeland T., Koller T., Murrin J., "*Valuation: Measuring and Managing the Value of Companies* ", John Wiley & Sons Inc., New York, 1990.

2. Copeland T.E., Keenan P.T., "*Making Real Options Real* ", The Mckinsey Quarterly, 1998, No.3

3. Hull J.C., "*Options, Futures and Other Derivatives* ", Prentice Hall of India Private Limited, New Delhi, 3rd Edition, 1998.

4. Damodaran A., "*Investment Valuation* ", John Wiley & Sons Inc., New York, 1996.

5. Gemill Gordon "*Options Pricing: An International Perspective*" , McGraw-Hill Book Company, Europe, April 1996.

6. Copeland T.E., Keenan P.T., "*How much is Flexibility worth* ", The Mckinsey Quarterly, 1998, No.2.

7. Brealey, R.A., Myers S.C., "*Principles of Corporate Finance* ", McGraw-Hill Inc., New York, 1996

8. Black F., Scholes M.S.*, "The valuation of options contracts and a test of market efficiency*", Journal of Finance, May, 1972

Appendix-1 Relations to Evolve Term Structure

Step-I

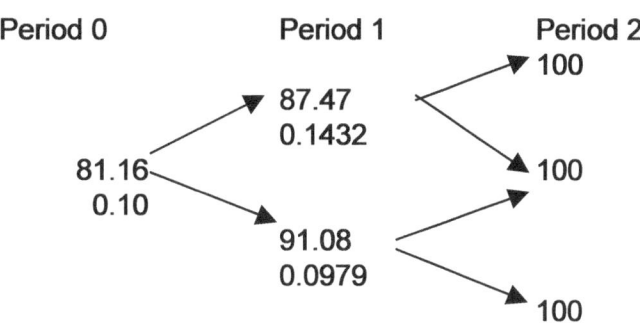

| Period 0 | Period 1 | Period 2 |

	87.47	100
	0.1432	
81.16		100
0.10		
	91.08	100
	0.0979	
		100

Relation 1	$81.16=(0.5S_u+0.5S_d)/1.10$
Relation 2	$0.5\text{Log}(r_u/r_d)=0.19$
Relation 3	$S_u=100/(1+r_u)$
Relation 4	$S_d=100/1+r_d)$
Substituting Relation 3 and Relation 4 in Relation 1	
Relation 5	$81.16*1.10=0.5(100/(1+r_u))+0.5(100(1+r_d))$

Solving Relation 2 and Relation 5	
r_u	0.1432
r_d	0.0979
S_u	87.47
S_d	91.08

Step II

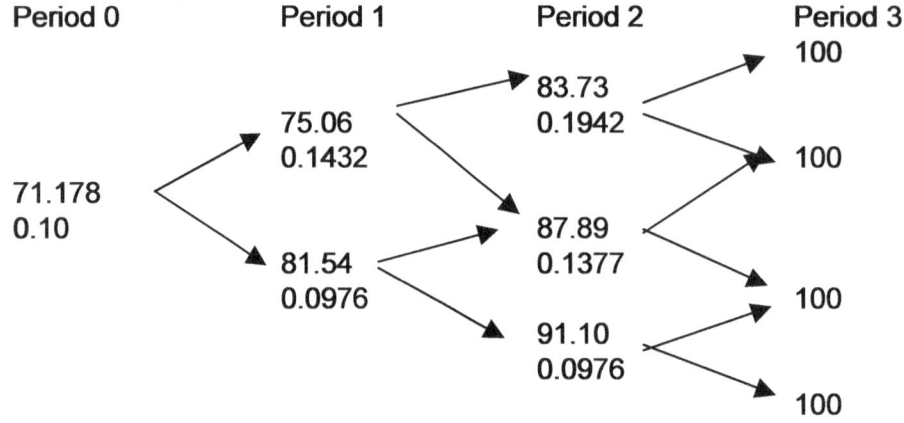

Relation 1	$71.178 = (0.5S_u + 0.5S_d)/1.10$
Relation 2	$0.5\log(r_{uu}/r_{ud}) = 0.18$
Relation 3	$r_{dd} = r_{ud}{}^\wedge 2/r_{uu}$
Relation 4	$S_{uu} = 100/(1+r_{uu})$
Relation 5	$S_{ud} = 100/(1+r_{ud})$
Relation 6	$S_{dd} = 100/(1+r_{dd})$
Relation 7	$S_u = (0.5S_{uu} + 0.5S_{ud})/(1+r_u)$
Relation 8	$S_d = (0.5S_{ud} + .05S_{dd})/(1+r_d)$

r_{uu}	0.1942
r_{ud}	0.1377
r_{dd}	0.0976
S_{uu}	83.73
S_{ud}	87.89
S_{dd}	91.1

Real Option Valuation via Stochastic Optimization

Markku Kallio

Helsinki School of Economics

April 8, 2002

Abstract

A multi-stage stochastic optimization model is presented to aid valuation of real as well as financial options. An option is specified by a compact set of alternative cash flow streams over the scenario tree of the stochastic program. The allowable choices in the model are specified by the set of option cash flows as well as by investments in competing assets, such as publicly traded financial instruments. Given preferences of the investor, the model determines simultaneously an optimal strategy for utilizing the option as well as an optimal investment strategy for competing assets. As suggested by Nau and McCardle, option value is determined as the buying price, the maximum price which the investor is willing to pay for the option. Such value is consistent with arbitrage theory of financial derivatives. If certain restrictions apply on preferences of the investor, then the option valuation results from computations of dynamic programming recursions.

Keywords: stochastic programming, real options, portfolio optimization

1. Introduction

Consider an investor planning to set up a new business or to expand its existing businesses. For example, a company may face a choice and timing of investments in a given set of potential technologies with uncertain prices of variable inputs and final outputs. To evaluate investment alternatives as real options we consider simultaneously investments in competing assets. For example, we may consider investments in a risk free asset and in stock market. Our valuation principle for such real options is based on buying price analysis, which originates from works by Nau and McCardle [6] and by Smith and Nau [9]; for exposition see also Luenberger [4]. We present some new results as well as their relationship to existing ones in [6] and [9]. For alternative approaches for real option valuation, see for instance, Copeland and Antikarov [2], Dixit and Pindyck [3] and Trigeorgis [10].

2. General option valuation

In this section we develop an option valuation procedure based on stochastic optimization. For this purpose, we begin by formulating a suitable optimization model. Thereafter we study properties for such models, and show that valuation based on buying price analysis is consistent with risk neutral valuation. General valuation procedures are discussed at the end on this section.

We employ a discrete time approach, where periods are defined by times $t = 0, 1, \ldots, T$. Hence, the time horizon is subdivided into T periods. An index $t > 0$ also refers to a period between time $t - 1$ and t. The duration in years between t and $t - 1$ may depend on t.

Uncertainty in our analysis may concern, for example, market prices of inputs and outputs of potential technologies in consideration as well as stock market prices of competing assets. Realizations of uncertainties over time are depicted by a scenario tree. Let k denote a node of the scenario tree with $k = 0$ referring to the root. Let k_- denote the predecessor of node k, for $k \neq 0$, and let K be the set of terminal nodes. Node k appears at time $t_k \in \{0, 1, 2, \ldots, T\}$. For the root, $t_0 = 0$, and for terminal nodes $k \in K$, $t_k = T$. For $k \neq 0$, we assume $t_{k_-} = t_k - 1$ for the predecessor node k_-. For $k \notin K$, let J_k denote the set of successor nodes of k. Hence, for all $j \in J_k$, we have $j_- = k$. The probability of attaining node k is π_k with $\pi_k > 0$, for all k.

A real option specifies a set of possible actions, such as technological investment and production strategy, for instance. A choice of actions yields a cash flow f_k, for all k. We denote the stochastic cash flow by $f = (f_k)$ and define a real option by a set F of feasible choices f. A charge denoted by V for such option is applied at time $t = 0$. We aim to determine an option value V as the maximum price the firm is willing to pay for the option.

Assumption A0. The set F is nonempty, closed and bounded.

Consider finitely many financial instruments including a risk free asset. For all such assets, let $S_{0,t}$ denote the vector of gross returns from time 0 to time t. For the risk free asset, gross return $R_{0,t}$ is a component of $S_{0,t}$. For times t and t' with $t < t'$, let $R_{t,t'} = R_{0,t'}/R_{0,t}$ be the risk free (forward) return from time t to t'. For $k \neq 0$, the

1

risk free return in period t_k is $R_k = R_{t_{k_-}, t_k}$. Similarly in period t_k, the vector of gross returns is denoted by S_k, the realization observed at node k.

Let vector s_k denote monetary amounts invested in financial assets at node k so that the investment expenditure at node k is $e^T s_k$, where $e^T = (1, 1, \ldots, 1)$. Let $s = (s_k)$ denote the financial investment strategy containing a subvector s_k for all k with $s_k = 0$, for terminal nodes $k \in K$. Transaction costs are excluded from consideration.

There is a private exogenous cash flow c_k, for each node k. For $k = 0$, c_0 is the initial cash. Such cash flows c_k may relate, for instance, to capacity existing before time $t = 0$. For each node k, we define an endogenous cash flow d_k, to be paid at node k. We call d_k dividend although other interpretations are possible. For instance, if d_k is negative, it may be interpreted as an increase in equity capital. In other cases it may be interpreted as consumption or wealth at the end of the horizon. For simplicity, taxes are not considered in our model. The firm is a risk-averse expected utility maximizer [7]. For practical purposes, we assume a separable utility function as follows:

Assumption A1. The utility function is separable, given by $u(d_0, d_1, \ldots, d_T) = \sum_t u_t(d_t)$. Let $\Phi \subset \{0, 1, \ldots, T\}$ such that $T \in \Phi$. For all $t \in \Phi$, assume $u_t(d_t) \in (-\infty, +\infty)$, $u'_t(d_t) > 0$ and $u''_t(d_t) < 0$, for all d_t, with $\lim_{d_t \to -\infty} u'_t(d_t) = \infty$. For all $t \notin \Phi$, define $u_t(d_t) = 0$, for $d_t \geq 0$, and $u_t(d_t) = -\infty$, for $d_t < 0$.

Hence, in some valuation cases we may account for d_t in some periods only, for instance, if the utility only depends on d_T, terminal wealth. Terminal dividend (wealth) always counts. Furthermore, if $t \in \Phi$, then u_t is strictly increasing and concave such that marginal utility increases without limit as d_t decreases.

The notation is summarized as follows:

f_k	option cash flow
F	feasibility set for option cash flow
V	option charge
c_k	private exogenous cash flow
d_k	dividend
s_k	financial investment vector
S_k	financial investment return vector
π_k	node probability

Given any V and $f = (f_k)$, consider the problem of finding strategies for financial investments $s = (s_k)$ and dividends $d = (d_k)$ to

$$\max \sum_k \pi_k u_{t_k}(d_k) \tag{2.1}$$

s.t.

$$e^T s_k - S_k^T s_{k_-} + d_k = f_k + c_k \quad \forall \ k \neq 0 \tag{2.2}$$

$$e^T s_0 + d_0 = f_0 + c_0 - V \tag{2.3}$$

$$s_k = 0 \quad \forall \ k \in K \tag{2.4}$$

We assume no arbitrage opportunities exist in the scenario tree. Formally, this is stated as follows

2

Assumption A2. For any $s = (s_k)$ and $d = (d_k)$, such that $s_k = 0$, for $k \in K$, $d_k = S_k^T s_{k_-} - e^T s_k$, for $k \neq 0$, and $d_0 = -e^T s_0$, $d \geq 0$ implies $d = 0$.

If A2 holds, then return vectors S_k are nonnegative, for all k, Furthermore, A2 with Stiemke's theorem of the alternative [5] implies, that for all k', there exist node prices $y_{k'} > 0$ such that

$$y_k e = \sum_{j \in J_k} y_j S_j \tag{2.5}$$

for $k \notin K$. Given a risk free asset exists, equivalently with the latter statement, there are risk neutral probabilities $q_j > 0$, for all $j \in J_k$, such that

$$Re = \sum_{j \in K_k} q_j S_j \tag{2.6}$$

where R is the risk free return over the period immediately succeeding node k. Finally, s and d in A2 constitute a homogenous solution for (2.2)-(2.4), for which $d \neq 0$ implies $d_k < 0$, for some k.

Lemma 1. *For problem (2.1)-(2.4), assuming A0 - A2,*

(i) an optimal solution exists,

(ii) optimal multipliers $\theta = (\theta_k)$ and θ_0 for (2.2) and (2.3), respectively, exist and they are strictly positive;

(iii) the optimal objective function value $u(V, f)$ is concave in f and V, strictly increasing in f and strictly decreasing in V; furthermore, $(-\theta_0, \theta)$ is a subgradient of $u(V, f)$ at (V, f);

(iv) for any V, an optimal solution $\hat{f} = \hat{f}(V) \in F$ exists for the problem

$$\max_{f \in F} u(V, f) \tag{2.7}$$

The optimal value $\hat{u}(V) = u(V, \hat{f}(V))$ is strictly decreasing and continuous in V, and

$$\lim_{V \to +\infty} \hat{u}(V) < u(0, 0) < \lim_{V \to -\infty} \hat{u}(V). \tag{2.8}$$

Proof: Clearly, the objective function in (2.1) is continuous, and a feasible solution $s = s^0$, $d = d^0$ exists for (2.1)-(2.4) with an objective function value denoted by u^0. Hence, to show (i), it suffices to to show that the closed and convex set D of feasible solutions $d \in D$ for which the objective function value is at least u^0, is bounded. Assume to the contrary, that D is not bounded. Then, for some vector $\delta^d = (\delta_k^d) \neq 0$, $d(\lambda) = d^0 + \lambda \delta^d \in D$, for all $\lambda \geq 0$. Furthermore, δ^d together with some vector $\delta^s = (\delta_k^s)$ constitute a homogenous solution for (2.2)-(2.4). Hence, A2 implies $\delta_k^d < 0$, for some k. Denote the objective function value along the ray by $u(\lambda)$. Assumption A1 implies that $t_k \notin \Phi$ and $u(\lambda)$ is concave in λ. The derivative is $u'(\lambda) = \sum_k \pi_k \delta_k^d u'_{t_k}(d_k(\lambda))$. Letting $\lambda \to \infty$, then by A1, u'_{t_k} remains bounded if $\delta_k^d \geq 0$, and it approaches ∞, if $\delta_k^d < 0$. Hence, $\delta_k^d < 0$, for some k, implies $\lim_{\lambda \to \infty} u'(\lambda) = -\infty$, and $u(\lambda)$ decreases without limit as λ increases so that for large enough λ, $d(\lambda) \notin D$. This is a contradiction, wherefore D is bounded. Hence, (i) follows, and we denote the optimal objective function value by $u(V, f)$.

Given (i) and A1, existence of optimal multipliers follows from Theorem 28.2 by Rockafellar [8]. We show the rest of (ii) by induction. Because $T \in \Phi$, optimality

3

conditions for d_k imply $\theta_k = \pi_k u'_T > 0$, for all terminal nodes $k \in K$. Next, consider any node k, such that $\theta_j > 0$, for all successor nodes $j \in J_k$. Optimality conditions for s_k require $\theta_k e = \sum_{j \in J_k} \theta_j S_j$, which together with A2 and (2.5) implies $\theta_k > 0$.

Assertion (iii) follows directly from (ii) and standard theory of convex optimization; see Rockafellar [8].

For assertion (iv), it follows from A0 and (iii) that an optimal solution $\hat{f} \in F$ for (2.7) exists. Next, let $V_1 < V_2$, with optimal choices $f_1 = \hat{f}(V_1)$ and $f_2 = \hat{f}(V_2)$. Suppose $\hat{u}(V_1) \leq \hat{u}(V_2)$ Then $\hat{u}(V_1) \leq u(V_2, f_2) < u(V_1, f_2)$, which is a contradiction. Hence, $\hat{u}(V)$ is strictly decreasing in V.

We now show continuity of $\hat{u}$. First, the epigraph of $\hat{u}(V)$ is closed, because it is the intersection of epigraphs of $u(V, f)$ which are closed, for all $f \in F$. Hence, $\hat{u}$ is lower semicontinuous. Second, consider the hypograph H of $\hat{u}(V)$, and a convergent sequence $\{u_n, V_n\} \subset H$ with $(u_n, V_n) \to (\bar{u}, \bar{V})$. Define $f_n = \hat{f}(V_n) \in F$. Then $\hat{u}(V_n) = u(V_n, f_n) \geq u_n$, for all n. By A0, there is a convergent subsequence $f_{n_i} \to \bar{f} \in F$, and $\hat{u}(\bar{V}) \geq u(\bar{V}, \bar{f}) \geq \bar{u}$, by continuity of $u(V, f)$. Hence, H is closed, and $\hat{u}$ is also upper semicontinuous. Consequently, $\hat{u}$ is continuous.

To prove the right side of (2.8), we denote by d^* an optimal dividend with $V = 0$ and $f = 0$, and proceed by picking V small enough such that risk free investments alone provide dividend $d \geq d^*$. Define $\alpha = \max_k d_k^*$, and for any $f \in F$, $\beta = \min_k(f_k + c_k)$. Then for any node k, risk free investment of $(\alpha - \beta)/R_{0,t_k}$ from time zero to time t_k yields at k a dividend $d_k \geq d_k^*$. Hence, for $V = -\sum_t (\alpha - \beta)/R_{0,t}$, we have $u(0,0) \leq u(V, f) \leq \hat{u}(V)$. The right side of (2.8) follows, as $\hat{u}(V)$ is strictly increasing in $-V$.

To show the left side in (2.8), define $f^* = (f_k^*)$, such that $f_k^* = \max_{f \in F} \max_k f_k$, for all k. Then A1 implies $u(V, f^*) \geq \hat{u}(V)$, for all V, and $\lim_{V \to +\infty} u(V, f^*) = -\infty$ by (iii). Hence, $\lim_{V \to +\infty} \hat{u}(V) = -\infty > u(0,0)$. $\square$

We also consider the problem (2.1)-(2.4) without the option. Then there is no cash flow from the real option and there is no option charge. Hence, in this case $f = 0$ and $V = 0$, and the problem represents optimal allocation of exogenous cash flow (c_k) alone. For this problem, we denote the optimal objective function value by $u^* = u(0,0)$. We value the option by indifference so that the option charge $V = \hat{V}$ is such that $\hat{u}(\hat{V}) = u^*$. In such a case the charge $\hat{V}$ is the buying price; i.e., $\hat{V}$ is the maximal price the investor is willing to pay for the option. Similarly by indifference, we may determine the selling price of an option; i.e. the minimum price at which one is willing to sell the option.

Theorem 1. *Assuming A0 - A2, there is a unique $\hat{V}$ with an optimal choice $\hat{f} = (\hat{f}_k) \in F$ for (2.7), such that $\hat{u}(\hat{V}) = u^*$. Furthermore, there exist node prices $y_{k'} > 0$, for all k', such that $y_k e = \sum_{j \in J_k} y_j S_j$, for all $k \notin K$, and*

$$\hat{V} = \sum_k y_k \hat{f}_k \tag{2.9}$$

in other words $\hat{V}$ is obtained by risk neutral valuation.

Proof: Lemma 1 (iv) implies a unique $\hat{V}$ such that $\hat{u}(\hat{V}) = u^*$. Furthermore, the optimum in (2.7), for $V = \hat{V}$, is attained with some $\hat{f} \in F$.

4

For $V = \hat{V}$ and $f = \hat{f}$, denote the optimal multipliers by $\hat{\theta}_k$ and $\hat{\theta}_0$, for (2.2) and (2.3), respectively. Similarly, for $V = 0$ and $f = 0$, denote the optimal multipliers by θ_k^* and θ_0^*. Then $\hat{u}(\hat{V}) = u^*$ and Lemma 1 (iii) imply

$$\hat{\theta}_0 \hat{V} - \sum_k \hat{\theta}_k \hat{f}_k \geq 0 \tag{2.10}$$

and

$$-\theta_0^* \hat{V} + \sum_k \theta_k^* \hat{f}_k \geq 0 \tag{2.11}$$

By Lemma 1 (ii), $\hat{\theta}_k > 0$, and optimality conditions imply $\hat{\theta}_k e = \sum_{j \in J_k} \hat{\theta}_j S_j$, for all $k \notin K$. Hence, $\hat{y}_k = \hat{\theta}_k / \hat{\theta}_0 > 0$ yield node prices for all k. Similarly, $y_k^* = \theta_k^* / \theta_0^* > 0$ yield node prices for all k. Hence, dividing inequality (2.10) by $\hat{\theta}_0$ and (2.11) by θ_0^*, yields

$$\sum_k y_k^* \hat{f}_k \geq \hat{V} \geq \sum_k \hat{y}_k \hat{f}_k$$

Hence, there is a convex combination of (y_k^*) and $(\hat{y}_k)$, denoted by (y_k), which satisfies (2.9). $\square$

In general, an approach for determining the option value is to employ a stochastic model for uncertain parameters first, and to generate the scenario tree thereafter, for example, by Monte Carlo simulation. Valuation of the option may follow from a search for $V = \hat{V}$, for instance, by trial and error, to satisfy $\hat{u}(\hat{V}) = u^*$. For computations, we deal with two cases A and B as follows. In *Case A* we consider investments in the real option F and in the financial assets. In this case, both financial investments, dividends and the choice $f \in F$ are simultaneously optimized, while the cash flow at the root node is reduced by an option charge V. The resulting problem is given by (2.1)-(2.4) and (2.7). In *Case B* we exclude the real option and the option charge so that the problem is given by (2.1)-(2.4) with $V = 0$ and $f = 0$. In case of indifference, the optimal objective function values are equal in both cases. Note, that the problems in *Case A* and *Case B* are convex optimization problems, if F is a convex set.

If there is a unique set of node prices $y_k > 0$, with $y_0 = 1$ for the root node, satisfying (2.5), then the valuation problem simplifies. In this case the market is complete so that any cash flow (f_k) can be replicated employing some investment strategy for financial assets. We may then begin by computing the set of prices y_k, with $y_0 = 1$. Thereafter, the value of any $f \in F$ is given by $\sum_k y_k f_k$. Hence, we have $\hat{V} = \max_{f \in F} \sum_k y_k f_k$, which clearly exists, for F nonempty and compact.

3.   Restricted preferences

In this section, we specialize to a case where dynamic programming by Bellman [1] is applicable. First, we discuss the preference restrictions, similar to those employed by Smith and Nau [9]. Thereafter, basic results are presented characterizing solutions of our option valuation problems. Finally, we discuss several alternatives of applying dynamic programming. For this aim, in addition to assumption A1 for the utility function, we consider exponential functions u_t as follows.

Assumption A3. In addition to A1, assume for all $t \in \Phi$, $u_t(d_t) = -\kappa_t \exp(-d_t / \gamma_t)$, where weight $\kappa_t > 0$ and risk tolerance $\gamma_t > 0$.

5

Employing risk free return $R_{t,t'}$ from time t to time t', define

$$\hat{\gamma}_t = \sum_{t' \geq t} \gamma_{t'} / R_{t,t'} \tag{3.1}$$

for $t = 0, 1, \ldots, T$. From (3.1) we obtain

$$\hat{\gamma}_{t-1} = \gamma_{t-1} + \hat{\gamma}_t / R_{t-1,t} \tag{3.2}$$

Lemma 2. *If A0 - A3 hold, then for the optimal objective function value in (2.7) , and for a decrease Δ in the option charge V, we have*

$$\hat{u}(V - \Delta) = \hat{u}(V) \exp(-\Delta/\hat{\gamma}_0). \tag{3.3}$$

Furthermore, an optimal choice $\hat{f} \in F$ in (2.7) is independent of the option charge V.

Proof: Given an option charge V, by Lemma 1, an optimal solution $\hat{s}$, $\hat{d}$ and $\hat{f}$ exists for (2.1)-(2.7). With $f = \hat{f}$ in (2.1)-(2.4), denote the optimal multipliers by $\hat{\theta}_k$ and $\hat{\theta}_0$, for (2.2) and (2.3), respectively.

Next, consider the revised problem (2.1)-(2.4) with option charge $V - \Delta$ and with $f = \hat{f}$. For this convex optimization problem, a feasible solution is obtained by adjustment in the initial solutions $\hat{s}$ and $\hat{d}$ as follows. For all $t \in \Phi$, invest in the risk free asset at time $t = 0$ until time t an amount Δ_t such that $\sum_t \Delta_t = \Delta$. For $t \notin \Phi$, define $\Delta_t = 0$. Thereby, dividend $\hat{d}_k$ is incremented by $R_{0,t_k} \Delta_{t_k}$, for all k. Additionally, for some α, we require $\exp(-R_{0,t} \Delta_t / \gamma_t) = \alpha$, for all $t \in \Phi$. Solving for α yields $\alpha = \exp(-\Delta/\hat{\gamma}_0)$.

For the resulting feasible solution of the revised problem, the objective function value is $\alpha u(V, \hat{f})$. Thereby, in comparison with the original problem, the gradient of the objective function is multiplied by α. Hence, multiplying the optimal multipliers $\hat{\theta}_k$ and $\hat{\theta}_0$ by α, we end up with a Karush-Kuhn-Tucker point for the revised problem. Therefore, the optimal objective function value satisfies

$$u(V - \Delta, \hat{f}) = \alpha u(V, \hat{f}). \tag{3.4}$$

Similarly, for any $f \in F$, $u(V - \Delta, f) = \alpha u(V, f)$. Hence, $\hat{f}$ is optimal for (2.7) with option charge $V - \Delta$, and $\hat{u}(V - \Delta) = \alpha \hat{u}(V)$. $\square$

For each node k of the scenario tree, a subtree with root k is defined by all nodes and arcs from time $t = t_k$ to T succeeding the node k. In the sequel, Lemma 2 is employed also for optimization over subtrees of the entire scenario tree. Of course, in place of $\hat{\gamma}_0$ we now have $\hat{\gamma}_t$ in (3.3). A decrease Δ in option charge can also be interpreted as an increase Δ in the exogenous cash flow at that node. Indeed, we employ Lemma 2 also to account for such variations in cash flow at the root k of the subtree.

Valuation of the option under A0 - A3 can be based on Lemma 2. To find the option value $\hat{V}$, we solve (2.1)-(2.4) and (2.7) first with option charge $V = 0$, and obtain an optimal objective function value $\hat{u}(0)$. Then, given the optimal value $u^* = u(0,0)$, by indifference and by Lemma 2, we have $u^* = \hat{u}(\hat{V}) = \hat{u}(0) \exp(\hat{V}/\hat{\gamma}_0)$ which yields

Theorem 2. *If A0 - A3 hold, then the option value is given by*

$$\hat{V} = \hat{\gamma}_0 \log[\, u^*/\hat{u}(0)\,]. \tag{3.5}$$

Similarly, the selling price V^*, the smallest price at which the firm is willing to sell the real option, satisfies $u(-V^*, 0) = \hat{u}(0)$ by indifference. Using Lemma 2, this yields $\exp(-V^*/\gamma_0)u^* = \hat{u}(0)$ so that $V^* = \hat{V}$. Hence, under assumptions A0 - A3, the buying price and selling price are equal.

6

3.1 Valuation of a stochastic cash flow

We now develop a dynamic programming recursion for valuation of any single choice $f \in F$. Thereafter, the option value is obtained by optimizing over $f \in F$. In this subsection, we begin with a dynamic programming formulation, which employs relatively simple optimization problems. Thereafter, assuming further restrictions on market, we discuss dynamic programing, where analytical solutions are available for recursions.

Consider a node k of the scenario tree with $t_k = t - 1$ so that the successor nodes of k appear at time t. Denote risk free return over period t by R. Recall that J_k is the set of successor nodes j of node k and S_j is the single period return vector for node j. Let $p_j = \pi_j/\pi_k$ be the conditional probability for node $j \in J_k$ given k.

Consider the problem (2.1)-(2.4) with option charge $V = 0$. Similarly, for a subtree with root node k and nodes l in the subtree, consider the problem

$$\max \sum_l (\pi_l/\pi_k)\, u_{t_l}(d_l)$$

s.t.

$$e^T s_l - S_l^T s_{l_-} + d_l = f_l + c_l \qquad \forall\ l \neq k$$

$$e^T s_k + d_k = f_k + c_k$$

$$s_l = 0 \qquad \forall\ l \in K$$

In the objective function, π_l/π_k is the conditional probability for node l, given root node k of the subtree. Let $\hat{u}_k$ denote the optimal objective function value of this problem.

3.1.1 Valuation by dynamic programming

For terminal nodes we have

$$\hat{u}_k = u_T(c_k + f_k) \qquad \forall\ k \in K. \tag{3.6}$$

For other nodes k with $t_k = t - 1 < T$, we employ leveling in dynamic programming; see Luenberger [4]. Let d be the dividend and s the vector of financial investments at node k such that $e^T s + d = f_k + c_k$. Then application of Lemma 2 in each successor node $j \in J_k$ and optimization over d and s yields

$$\hat{u}_k = \max_{e^T s + d = f_k + c_k} \left\{ u_{t-1}(d) + \sum_{j \in J_k} p_j \hat{u}_j \exp(-S_j^T s/\hat{\gamma}_t) \right\} \tag{3.7}$$

where $d = 0$ is optimal, if $t - 1 \notin \Phi$.

For the case without an option we have $f = 0$. In this case, denote the optimal objective function value in (3.6)-(3.7) by u_k^* instead of $\hat{u}_k$. Then, at the root node $k = 0$, using notation of Lemma 1, $\hat{u}_0 = u(0, f)$ and $u_0^* = u(0, 0)$. Hence, the option value $\hat{V}$ is obtained by Theorem 2 as

$$\hat{V} = \hat{\gamma}_0 \log(u_0^*/\hat{u}_0). \tag{3.8}$$

7

3.1.2 Partially complete market

For further specialization in dynamic programming, we subdivide set J_k into disjoint subsets J_{ik}, for $i \in I_k$, so that $S_j = S_i$ for all $j \in J_{ik}$. At node k, p_i denotes the probability for S_i, and given $i \in I_k$, the conditional probability for node $j \in J_{ik}$ is r_{ij} so that $p_j = p_i r_{ij}$. Under assumption A2, by (2.6), there are risk neutral probabilities $q_i > 0$, $i \in I_k$, such that $Re = \sum_i q_i S_i$, where $e^T = (1, 1, \ldots, 1)$. For dynamic programming recursions, we employ a further assumption concerning completeness of financial market. Subsequently, we obtain an auxiliary result, which then yields analytical expressions for optimal objective function values.

Assumption A4. For all nodes k, risk neutral probabilities $q_i > 0$, for $i \in I_k$, are unique.

Lemma 3. *Consider a single period starting at node $k \notin K$. If d is a constant, $\gamma > 0$, $h_i > 0$, for all $i \in I_k$, are positive scalars, and $e^T = (1, 1, \ldots, 1)$, then A4 implies*

$$\min_s \{ \sum_i h_i \exp(-S_i^T s/\gamma) \mid e^T s = d \} = \exp(-Rd/\gamma)\Pi_i(h_i/q_i)^{q_i}. \qquad (3.9)$$

Proof: By Lemma 1 (i), an optimal solution s^* for the problem on the left side of (3.9) exists. Denote $u_i = \exp(-S_i^T s^*/\gamma)$ so that $\bar{u} = \sum_i h_i u_i$ is the optimal objective function value. Optimality condition and existence of a risk free asset, yield $Re = \sum_i q_i' S_i$ and $1 = \sum_i q_i'$, where $q_i' = h_i u_i/\bar{u} > 0$. Hence, $q_i' = q_i$ is the unique risk neutral probability, for all $i \in I_k$. Definitions of u_i and q_i' yield $\log u_i = -S_i^T s^*/\gamma = \log(\bar{u} q_i/h_i)$. Multiplying the latter equality by q_i and summing over $i \in I_k$ yields $-\sum_i q_i S_i s^*/\gamma = \sum_i q_i \log(\bar{u} q_i/h_i)$, where the left side is $-Rd/\gamma$, because $\sum_i q_i S_i = Re$ and $e^T s^* = d$. Hence, we obtain $-Rd/\gamma = \sum_i q_i \log \bar{u} + \sum_i q_i \log(q_i/h_i) = \log \bar{u} - \log \Pi_i(h_i/q_i)^{q_i}$, which yields (3.9). $\square$

Equation (3.7) is now rewritten as

$$\hat{u}_k = \max_{e^T s + d = f_k + c_k} \{ u_{t-1}(d) + \sum_{i \in I_k} p_i \sum_{j \in J_{ik}} r_{ij} \hat{u}_j \exp(-S_i^T s/\hat{\gamma}_t) \}$$

where $d = 0$ is optimal, for $t - 1 \notin \Phi$. Consequently, using Lemma 3,

$$\hat{u}_k = \max_d \{ u_{t-1}(d) - \exp[R(d - f_k - c_k)/\hat{\gamma}_t]\Pi_{i \in I_k}(-p_i \hat{u}_i/q_i)^{q_i} \} \qquad (3.10)$$

where

$$\hat{u}_i = \sum_{j \in J_{ik}} r_{ij} \hat{u}_j \qquad (3.11)$$

If $t - 1 \in \Phi$, then optimization over d in (3.10) with (3.2) yields, after lengthy but straightforward algebra,

$$\hat{\gamma}_{t-1} \log(-\hat{u}_k) = -c_k - f_k + \psi_{t-1} + (1/R) \sum_{i \in I_k} q_i \hat{\gamma}_t \log(-p_i \hat{u}_i/q_i) \qquad (3.12)$$

where

$$\psi_{t-1} = \gamma_{t-1} \log(\kappa_{t-1}) - (\hat{\gamma}_t/R) \log(\hat{\gamma}_t/R) + \hat{\gamma}_{t-1} \log \hat{\gamma}_{t-1} - \gamma_{t-1} \log \gamma_{t-1} \qquad (3.13)$$

is independent of option cash flow f. If $t - 1 \notin \Phi$, then $d = 0$ and $\hat{\gamma}_{t-1} = \hat{\gamma}_t/R$ by (3.2), so that (3.10) implies (3.12) with $\psi_{t-1} = 0$.

Again, for the case without an option, denote the optimal objective function value over the subtree by u_k^* in (3.6) and (3.12). Then, at the root node $k = 0$, we obtain the option value $\hat{V}$ from (3.8).

3.1.3 Buying price recursion

Alternatively, we may apply dynamic programming for valuation of cash flow f employing a buying price v_k instead of optimal value $\hat{u}_k$, for each node k. For cash flow f, in the subtree with root node k we define the buying price v_k by indifference. For terminal nodes $k \in K$, we have $v_k = f_k$. In general, using the notation above and Lemma 2, we have

$$\hat{u}_k = u_k^* \exp(-v_k/\hat{\gamma}_{t-1}) \tag{3.14}$$

so that $v_k = \hat{\gamma}_{t-1} \log(u_k^*/\hat{u}_k)$. This is rewritten using (3.11)-(3.12) both for u_k^* and $\hat{u}_k$, after elementary algebraic steps, as follows

$$v_k = f_k - (1/R) \sum_{i \in I_k} q_i \hat{\gamma}_t \log\Big[\sum_{j \in J_{ik}} r_{ij} \hat{u}_j / \sum_{j \in J_{ik}} r_{ij} u_j^* \Big] \tag{3.15}$$

Definition (3.14) applied to node $j \in J_{ik}$ yields $\hat{u}_j = u_j^* \exp(-v_j/\hat{\gamma}_t)$. Then, for $i \in I_k$ and $j \in J_{ik}$, definition

$$r_{ij}^* = r_{ij} u_j^* / \sum_{j \in J_{ik}} r_{ij} u_j^* \tag{3.16}$$

and equation (3.15) yields the following result

Lemma 4. *If A1-A4 hold, then the value of a bounded cash flow f in a subtree with root node k is given by $v_k = f_k$, for $k \in K$, and for $k \notin K$, by recursion*

$$v_k = f_k + (1/R) \sum_{i \in I_k} q_i C_i \tag{3.17}$$

where

$$C_i = -\hat{\gamma}_t \log\Big[\sum_{j \in J_{ik}} r_{ij}^* \exp(-v_j/\hat{\gamma}_t) \Big]. \tag{3.18}$$

with probabilities r_{ij}^* defined by (3.16) and parameters $\hat{\gamma}_t$ by (3.1).

Here C_i in (3.18) represents certainty equivalent for prices v_j obtained with probabilities r_{ij}^*, when an exponential utility function with risk tolerance $\hat{\gamma}_t$ is employed. The notation is illustrated in Figure 2 with node k and four successor nodes. There are two distinct return vectors S_i, so that there are two elements i in the set I_k and two elements in each J_{ik}.

$t-1$	p_i	C_i	r_{ij}	t	v_{ij}	S_i
		C_1	r_{11}	•	v_{11}	S_1
	p_1	.				
v_k			r_{12}	•	v_{12}	S_1
•						
c_k			r_{21}	•	v_{21}	S_2
f_k	p_2	.				
		C_2	r_{22}	•	v_{22}	S_2

9

Figure 2. A node k at time $t-1$ and its successor nodes ij at time t.

Next, consider the special case where, for each node $k \notin K$, u_j^* is independent of j, for all $j \in J_{ik}$. Then $u_j^* = u_i^*$ in (3.11) and $r_{ij}^* = r_{ij}$ in (3.16), and (3.17) yields the roll back procedure by Smith and Nau [9]. In this case, u_k^* is unnecessary for recursion (3.17). For each node $j \in J_{ik}$, subsequent realizations of stock return and of exogenous cash flow are independent of j, but they may be dependent within any single time period. In real option analysis this may not be sufficient. It can be crucial to allow future development of stock prices depend on current price developments outside the financial market, for example, in case where the stock prices are correlated with price processes determining option cash flow. Also in cases where the real option refers to capacity expansion, the option cash flow f_k is likely to be correlated with exogenous cash flow c_k resulting from existing capacity.

3.2 Option valuation

We now turn to application of dynamic programming for solving simultaneously problems (2.1)-(2.4) and (2.7) for optimal s, d and f, given a fixed option charge $V = 0$. For this aim, we restrict the definition of the option cash flow $f \in F$, which now results from a multistage decision process as follows.

For each node k in the scenario tree, assume there is a finite number of possible choices to be made concerning the option. For such a choice, the path in the scenario tree from the root up to node k as well as all choices preceding node k are known. Hence, possible choices at node k depend on node k and the choice history denoted by l. Let H_{kl} define the set of feasible choices at k, given choice history l. Each element $h \in H_{kl}$ denotes the choice history l appended by a subsequent choice at node k. Thus, $h \in H_{kl}$ denotes a choice history at successor nodes $j \in J_k$, given choice history l at node k. We assume that return of financial assets and private exogenous cash flow are independent of all choices made.

For all k, option cash flow f_k depends on node k, on the history l prior to node k, as well as on the choice $h \in H_{kl}$. Hence, we denote $f_k = f_{kl}(h)$. Assuming that all such cash flows $f_k = f_{kl}(h)$ are finite, the finite set F of possible option cash flows $f = (f_k)$ is bounded. Consequently, assuming choice sets H_{kl} such that F is nonempty, then A0 holds.

3.2.1 Option valuation by dynamic programming

We carry out dynamic programming recursion in the tree with nodes kl appearing at time t_k. Assuming A0-A3, recursion for solving (2.1)-(2.4) and (2.7) is developed using results of Section 3.1. Let $\hat{u}_{kl}$ be the optimal expected utility over the subtree with root node kl. For terminal nodes $k \in K$, given choice history l,

$$\hat{u}_{kl} = u_T[c_k + \max_{h \in H_{kl}} f_{kl}(h)]. \tag{3.19}$$

For other nodes k with $t_k = t - 1 < T$, it follows from Lemma 2, that an optimal choice $h \in H_{kl}$ at node kl is independent of exogenous cash flow at node kl. As in Section 3.1, we employ leveling. The optimal value $\hat{u}_{kl}$ is then obtained in two stages.

10

First, for each choice $h \in H_{kl}$, we optimize dividend d and financial investments s with $e^T s + d = f_{kl}(h) + c_k$. Given choice $h \in H_{kl}$, we obtain the optimal expected utility

$$\hat{u}_{kl}(h) = \max_{e^T s + d = f_{kl}(h) + c_k} \{u_{t-1}(d) + \sum_{j \in J_k} p_j \hat{u}_{jh} \exp(-S_j^T s / \hat{\gamma}_t)\} \quad (3.20)$$

Again, $d = 0$ is optimal in (3.20), if $t_k \notin \Phi$. Second, optimization over choices $h \in H_{kl}$ yields

$$\hat{u}_{kl} = \max_{h \in H_{kl}} \hat{u}_{kl}(h). \quad (3.21)$$

Then by Theorem 1, given u_k^* in Section 3.1, we obtain the option value

$$\hat{V} = \hat{\gamma}_0 \log(u_0^* / \hat{u}_{00}), \quad (3.22)$$

where $k = 0$ refers to the root node and $l = 0$ to the void choice history at the root.

3.2.2 Option valuation under a partially complete market

Next, assume A0-A4 hold. Then with (3.12), for nodes kl with $t_k = t - 1 < T$, we obtain

$$\hat{\gamma}_{t-1} \log[-\hat{u}_{kl}(h)] = -c_k - f_{kl}(h) + \psi_{t-1} + (1/R) \sum_{i \in I_k} q_i \hat{\gamma}_t \log(-p_i \hat{u}_{ih} / q_i) \quad (3.23)$$

where ψ_{t-1} is given by (3.13), for $t_k \in \Phi$, and $\psi_{t-1} = 0$, for $t_k \notin \Phi$, and

$$\hat{u}_{ih} = \sum_{j \in J_{ik}} r_{ij} \hat{u}_{jh}$$

Then given $\hat{u}_{kl}$ in (3.23) and (3.21), and u_k^* in Section 3.1, we obtain the option value from (3.22).

Alternatively, let v_{kl} denote the buying price for option cash flow in a subtree with node kl as the root. Hence, by Lemma 2, $\hat{u}_{kl} = u_k^* \exp(-v_{kl} / \hat{\gamma}_{t-1})$. For node kl with $k \in K$, we obtain

$$v_{kl} = \max_{h \in H_{kl}} f_{kl}(h) \quad (3.24)$$

and for other nodes kl, using (3.15), recursively

$$v_{kl} = \max_{h \in H_{kl}} \{f_{kl}(h) - (1/R) \sum_{i \in I_k} q_i \hat{\gamma}_t \log[\sum_{j \in J_{ik}} r_{ij} \hat{u}_{jh} / \sum_{j \in J_{ik}} r_{ij} u_j^*]\} \quad (3.25)$$

where $\hat{u}_{jh} = u_j^* \exp(-v_{jh} / \hat{\gamma}_t)$. Hence, (3.25) yields the following result

Lemma 5. *If A0 - A4 hold, then the option value v_{kl} is given by (3.24), for kl with $k \in K$, and for other nodes kl, by recursion*

$$v_{kl} = \max_{h \in H_{kl}} \{f_{kl}(h) + (1/R) \sum_{i \in I_k} q_i C_{ih}\} \quad (3.26)$$

where

$$C_{ih} = -\hat{\gamma}_t \log[\sum_{j \in J_{ik}} r_{ij}^* \exp(-v_{jh} / \hat{\gamma}_t)] \quad (3.27)$$

with probabilities r_{ij}^* defined by (3.16) and parameters $\hat{\gamma}_t$ by (3.1).

Similarly as in Section 3.1, if $u_j^* = u_i^*$, for all $j \in J_{ik}$, then $r_{ij}^* = r_{ij}$. In this case, Theorem 3 results in the roll back procedure by Smith and Nau [9].

11

References

[1] Bellman, R., *Dynamic Programming*, Princeton University Press, 1957.

[2] Copeland, T. and V. Antikarov, *Real Options*, Texere, New York, 2001.

[3] Dixit, A.K. and R.S. Pindyck, *Investment under Uncertainty*, Princeton University Press, 1994.

[4] Luenberger, D.G., *Investment Science*, Oxford University Press, 1998.

[5] Mangasarian, O., *Nonlinear Programming*, McGraw-Hill Book Company, 1969.

[6] Nau, R.F. and K.F. McCardle, "Arbitrage, rationality and equilibrium," *Theory and Decision* 31 (1991) 199-240.

[7] von Neumann, J. and O. Morgenstern, *Theory of Games and Economic Behavior*, Princeton University Press, 1947.

[8] Rockafellar, R.T., *Convex Analysis*, Princeton University Press, 1970.

[9] Smith J.E. and R.F. Nau, "Valuing Risky Projects: Option Pricing Theory and Decision Analysis," *Management Science* 41 (1995) 795-816.

[10] Trigeorgis, L., *Real Options. Managerial Flexibility and Strategy in Resource Allocation*, The MIT Press, 1997.

12

DAY 2

Tuesday 07.05.

Chairman   Juha Paappanen

SESSION 1

09.00 – 09.30 Marie-Laure Guilherminet
"The decision of investment and its funding in an undergoing institutional environment: the case of nuclear equipment"

09.30 – 10.00 Marcus Dimpfel and René Algesheimer
"Real Options Theory and Broadcasting"

10.00 – 10.30 Rautaruukki Case

10.30 – 10.45 Morning coffee

SESSION 2

11.00 – 12.00 Keynote Speaker
Alexander Bukhvalov
"Application of Real Options to Strategic Management in Transition Economies"

12.00 – 13.00 Lunch

SESSION 3

13.00 – 14.00 Christer Carlsson and Robert Fullér
"Fuzzy Real Option Valuation – A Breakthrough Theory"

14.00 – 14.30 Mikael Collan
"Investment Decisions and Fuzzy Numbers"

14.30 – 15.00 Péter Majlender
"Optimal Timing for the Exercise of Real Options"

15.00 – 15.30 Markku Heikkilä
"User Interfaces and Knowledge Presentation in Real Options"

15.30 – 15.45 Afternoon coffee

SESSION 4

15.45 – 16.45 Keynote Speaker
Marco Guimaráes Dias, Petrobras, Brazil
"Overview of Real Options in Petroleum"

16.45. – 17.15 Francisco Alcaraz

17.15. – 17.45 Fortum Case

19.30 – *Evening program* – Banquet at the Castle Cellar

DAY 2

SESSION 1

Marie-Laure Guilleminet
"The Decision of Investment and its Funding in an
  Undergoing Institutional Environment: The case of
  Nuclear Equipment"

Marcus Dimpfel and René Algesheimer
"Real Options Theory and Broadcasting"

*Rautaruukki Case ⑪ **RAUTARUUKKI**

*No Material

The decision of investment and its funding in an undergoing institutionnal environment: the case of a nuclear equipment

Marie-Laure Guillerminet[*] 15th

March 2002

Abstract

This paper's objective is to estimate the consequences of the questioning of the independance hypothesis of investment and financing decisions, on the investment choice of a nuclear electricity producer in an European market which opens in the competition. This firm which can get into debt on the financial market, is subjected to tax and can go bankrupt. The firm takes into account its capital structure before accepting the project. We show that its totally irreversible investment opportunity becomes partially reversible when the investment and financing decisions are interdependent.

Keywords: deregulation, investment, real option, optimal financing, cost price

JEL Codes: D92, G32

[*]CREDEN-LASER, Correspondance: Creden, Université Montpellier I, UFR Sciences Economiques, Espace Richter, Avenue de la Mer, BP 9606, 34054 Montpellier cedex 1, France, Phone: +33 (0)4 67 15 83 32, E-mail: mlg@sceco.univ-montp1.fr

1

1 Introduction

The institutional context of the French electricity sector is marked by the European Directive of December 19th, 1996. In France the producer was previously integrated into a public natural monopoly. Now articles 4, 5 and 6 of the electric Directive open the generation sector to the independent power producers (IPPs). Forced furthermore to unbundle its activities, Electricity de France (EdF) is going to become an IPP. The Directive does not foresee the modification of the industrial structure which establishes the opening in competition. We envisage, in reference to the British case, the emergence of the European electricity pool, the existence of which will not necessarily be concretized. However if the pool is the retained industrial structure, it will not be created before the total opening of the generation sector, probably in 2006.

This report coincides with the question of the renewal of the power plants park foreseen by 2010, i.e. the end of the 900 and 1300 MW nuclear equipments' life expectancy, because the French production is 80% of nuclear origin. The case of the United Kingdom allows us to show that the privatization of the electricity producers, including nuclear power plants (NPPs), is possible during the competition implementation: British Energy[1], privatized in 1996, is a concrete example. We limit in this paper the privatization to the opening of the producer's capital and not to its increase: we speak in a general way about the firm's shareholders while it is at first about the State for EdF. Furthermore, the capital is supposed to consist only of common shares and bonds, that is the firm can finance its capital by equity and loan. So the French EdF firm which holds the NPPs will not keep necessarily its status of public establishment with industrial and commercial character.

What incidence can the capital opening of the electricity producer have on its decision to invest in an additional nuclear equipment?

The firm's environment is competitive and considered as uncertain. The envisaged specific capital is a based-load nuclear equipment which has a totally sunk cost: the firm cannot cut investment if the market conditions become unfavourable. On the other hand, the firm keeps some flexibility in the construction timing of this equipment which is approved by the regulator. These three characteristics of uncertainty, irreversibility and timing flexibility which characterize this project interact on the investment decision. They allow to estimate the investment opportunity of the firm as a real option, and more exactly as a call option because the equipment is totally irreversible.

The value of the investment project is thus calculated by the theory of real options by supposing the unlevered firm, what means admitting that the Modigliani-Miller[1958] hypothesis of separation of the investment and financing decisions is verified. Now, in finance, the trade-off theory questions this hypothesis (section 2). We will see that the interaction between the

---

[1] The generation park of British Energy does not however consist exclusively of NPPs, which furthermore were partially paid off before the privatization of the firm.

2

optimal investment and financing decisions is going to be translated by adding a put option to the call option with which is identified the investment opportunity: the project is not totally irreversible anymore. On the pool, the investment decision is not necessarily delayed anymore, as for an unlevered firm, with regard to that determined by the standard net present value (NPV) rule obtained in the monopoly (section 3). In conclusion, the levered investment is not totally irreversible anymore (section 4).

2 The characteristics of the nuclear investment and its funding structure

2.1 The nuclear project value by the theory of real options

2.1.1 The cost price evolution

The evolution of the institutional frame takes into account the decentralization of the electric generation policy which follows upon the opening of European sector in competition. It is schematized by the market modification from the regulated integrated monopoly to the competitive pool. The industrial structure is controlled by a regulator which acts for the collectivity and which exercises its influence on the firm's cost price evolution. This regulator has for objective productive and allocative efficiency and to do it, tries to implement a disputable market because the monopolistic cost-of-service regulation presents two main limits: the Averch-Johnson effect of overcapitalization and the financial inefficiency. We notice that the certain environment of the producer becomes uncertain. The only uncertainty source which we isolate is the cost price, parameter which the firm does not control partially and which nevertheless drives its investment decision. It reports the industrial structure and constitutes the project's equity.

Certain evolution in monopoly - The public firm has to fulfil Universal Service Obligations (USOs) which are imposed by the State, the market regulator, and which force it to decline its cost price to avoid any abuse of dominant position.

We thus suppose that the cost price movement, which characterizes a cost-of-service regulated monopolistic sector, is the following one:

$$\frac{dp}{p_t} = \bar{\mu}dt, \tag{1}$$

where the trend of the rate evolution $\bar{\mu}$ is negative, by definition ($\bar{\mu} \mathbf{6}\ 0$): the cost price lowers as the long term marginal cost decreases.

Uncertain evolution on the pool - The regulator supervises the firm on the pool (cf. the British example of the Electricity Pool of England and Wales, and since April 1st, 2001 of New Electricity Trading Arrangements) to maintain competition. The purpose of the deregulation which introduced competition into the generation sector, is to lower the cost price following the

3

long term marginal cost. The pool price is the same for any producer. It is established from the marginal cost of the incremental power plant called by the pool in the merit order bids system. The regulator fixes the pool's operative rules and so determines the decreasing cost price evolution which is imposed to the firm.

The cost price evolution which characterizes the pool follows a geometrical brownian motion, usual model of the uncertain variables evolution in finance:

$$\frac{dp}{p_t} = \bar{\mu}dt + \sigma dz, \tag{2}$$

where $\bar{\mu} \, \mathbf{6} \, 0$ as previously;

the volatility σ, which characterizes the risk of the firm's activity, is positive: $\sigma > 0$.

2.1.2 Definition of the various costs

Infinite life expectancy - The DIGEC[1997] estimates the duration of reactors' exploitation in a range of 30 at 50/60 years (40 years for N4 reactors) which we consider as infinite.

Standardized equipment - The composition of the nuclear park settled by EdF is such as its accumulated experience is estimated at 600 "reactors-years". The initial capital is known, even if the construction delays of a NPP are long. They are nevertheless supposed to be of 16 years for the first 900 MW REP in 6 years for N4 reactors, but remain superior to those of a gas combined cycle turbine (2 years). Because of this technology standardization, we are going to suppose that the construction delay is immediate, what means not taking into account the cost price evolution between the time of investment (and its financing) decision and that of the equipment's operation. By hypothesis, the initial capital[2], which includes construction, R&D and dismantling[3] costs, is then normalized to K.

The based-load operating - Because of the costs structure, the nuclear French equipment is competitive in based-load generation, i.e. intended to work 7000 at 8760 hours a year. We suppose that the production is constant, considering the avaibility rate of NPPs,and normalized to the unity. Furthermore, to restrict the uncertainty to the only electricity cost price variation, the uranium price, nevertheless uncertain, is supposed constant. In a similar way, the study of the DIGEC[1997] frames the evolution of the uranium price by two "low" price (20 USD/1b U308) and "height" (25 USD/1b U308) scenarios. These two hypotheses of fixed fuel price and constant production allow to normalize the production marginal cost to 0. The cash flows, generated by the investment project, can thus be reduced to the cost price.

---

[2] This latter remain fixed in a favorable environment. There was no accident since 1986, which would have been able to pull a costs increase, and if the nuclear can benefit from a premium further to the Kyoto agreements (from 1 till 10 December 1997), it will be known and thus normalized.

[3] We do not take into account very strong uncertainties on the last stages of the cycle, both in terms of feasibility and costs, as far as they concern a very distant future and a standardized equipment. These expenses represent only 2 % of the accumulated spendings of the French nuclear park.

4

The competition is not taken into account - The nuclear security standards were strengthened further to the accidents arisen on the reactors of Windscale in Great Britain from 7 till 12 October 1957, Three Mile Island in the United States on March 28th, 1979 and of Tchernobyl in Ukraine on April 26th, 1986. They maintain the nuclear electricity producer under the regulator's control, what eliminates any strategic effect on the project value (cf. Trigeorgis[1996, p. 134]). So the additional nuclear equipment is an opportunity to invest for EdF. Also, the NPPs begin to be competed in based-load only by the gas combined cycles turbines: we so suppose that the expansion of the firm's capital is total and that the initial capital cost is constant (cf. Dixit and Pindyck[1998]).

The totally irreversibility of the equipment - This irreversibility implies a sunk cost: the firm cannot cut investment if the market conditions become unfavourable and the market exit is made of the initial capital cost. The firm's capital is specific. Furthermore, if the firm's capital resale is possible, the potential buyers are in the same market conditions which urge the firm to sell its capital. They are all the less incited to invest that the price is close to the replacement cost. Thus even if the capital is not firm's specific, the industry's specificity of both capital and shocks make the investment irreversible (cf. Akerlof[1970]).

The firm has the opportunity to buy capital without having that to resell it in case of reversal of the market conditions: this irreversibility characteristic generates the real option consisted only of an call option.

2.1.3 Flexibility of the investment decision

The network industries' firms have to fulfil USOs. And so, in the electric sector, the State is the owner of the vertically integrated natural monopoly. By introducing the competition, the transposition of the European electricity Directive leaves unresolved the question of production capacities and energy choices for France. It foresees that the public transport system operator (SO) establishes, at least every other year and under control of the State, a multiannual projected balance of needs in production capacities. On the basis of this balance, the State plans the production investments, by granting exploitation licenses and if these last ones do not answer the fixed objectives, by proceeding to bids after refering to the opinion of the Electricity Regulation Committee and the transport SO. The regulator is in charge of organization and control of the USOs. On the other hand, the firm keeps the investment timing. It has the opportunity to invest in a nuclear equipment, and it can, at any time, decide to invest or to wait. It does not lose this opportunity if it decides to wait before invest: there is no competitive game against others firms. It is then a question for the firm of determining the optimal time of its investment, and of deciding on its financing.

These three characteristics of uncertainty, irreversibility and flexibility allow to estimate the project by means of the theory of real options introduced by Abel[1983], Pindyck[1988],

5

Caballero[1991], Smith[1994], Dixit[1995]. This literature on the investment decisions taken in uncertain environment assimilates the investment opportunity to an option. It integrates into the standard net present value (NPV) of the equipment the wait value of supplementary information about the future, in the fact that Trigeorgis[1996] called the expanded net present value (ENPV).

The NPV and the ENPV define respectively the investment thresholds in monopoly and on the pool. Indeed the firm can decide to realize the investment or to wait. It sees proposing the investment project in the initial period. It estimates it according to the NPV rule. If this project is not profitable, the firm does not hold it to the initial period. But, in the next period, the choice remains the same and the firm again has to decide to invest or to wait, having acquainted with the production cost which increased or decreased. This possibility of wait has some value, that of the real option.

The firm which invests, exercises the option which it holds. Finally, the choice terms of the unlevered firm answer the control u_t, $u_t \in \{$invest; wait$\}$:

- if the option is exercised at time t, it means that the net present cash flow of this investment $(p_t - K)e^{-rt}$ is positive, where p_t is the cash flow which is reduced to the cost price according to our hypotheses, K is the initial amount of the capital and r is the risk-free interest rate;

- if the option is not exercised at this time t, there is no production earning and this option value is zero.

Among three mentioned conditions, the literature concentrates on the investment irreversibility and so, the wait is cost-free. Further to Pindyck[1988], Abel and Eberly[1996, 1997], both Abel, Dixit, Eberly and Pindyck[1996], Dixit and Pindyck[1998] models analyzed the influence of the capital reversibility and expansion on the option premium Δ, which is defined as the difference between the ENPV and the NPV. *Any unlevered investment so defined by the real call is totally irreversible and expansible. It is characterized by a positive option premium:* $\Delta > 0$.

Our problem consists in showing the consequences of the questioning of the Modigliani-Miller[1958] hypothesis of the independence of investment and financing decisions on the option premium value Δ. Now Abel and al.[1996], Dixit and Pindyck[1998] put in motto that Δ is not always positive for a partially reversible investment.

2.2 Funding characteristics

2.2.1 The project financing due to the risks

The nuclear industry is confronted with a specific economic risk, even if this risk is diversifiable by the investors as proves it to us the privatization of British Energy[4] in 1996 in the United

---

[4] The generation park of British Energy does not consist exclusively of NPPs, which furthermore were partially paid off before the firm's privatization.

6

Kingdom. It includes the risks due to waste and to accidents possibility, which are partially solved by engeneering. Their coverage, identical in France and in the United Kingdom, advances the limited producer's responsibility in amount and in time. In case of the firm's or insurer's insolvency[5], the State assumes all or any of this liability[6]. The risks also concern the competitiveness of NPPs as based-load production means. The technological evolution makes possible the appearance of a new based-load offer. The pool reveals the firm's marginal cost by calling it in the merit order: there is thus a technological incitement to have the lowest marginal cost[7]. Finally the risks are due to the capital-intensive character and the irreversibility of the nuclear investment. Finally the cost price is guarantee-free, except in case of existence of long term purchase contracts, and the financing of a nuclear project is thus risky.

These risks are integrated into the economic calculation by means of their premiums. The financing possibility of these nuclear equipments is not then questioned anymore. We are going to concentrate on the sharing of the nuclear equipment's financing between equity and loan. We are not going to detail the loans' typology for the NPP's funding, but rather to determine the investment thresholds of the levered electricity producer. The risks are assumed for every investment project, the financing of which is made via a specific legal structure[8], the project-corporation (Lescoeur and Penz[1999]).

2.2.2 Interdependence between investment and financing decisions

The trade-off theory - The trade-off theory (Myers[1977]), considers two market imperfections, which are tax and bankruptcy. It is in continuation of the theory of Modigliani-Miller[1958], because it does not question the leverage. But it adds to leverage two effects due to debt, so that the firm's value varies according to the debt rate:

- the tax effect: the debt interests are considered as corporate expenditure and are deductible from the taxable profit;

- the bankruptcy[9] risk and the insolvent probability increase with the debt.

Its main result thus establishes the existence of an optimal debt rate which maximizes the project value: the financing and investment decisions are then interdependent. So the firm

---

[5] We shall use indifferently the term of insolvency, i.e. the state of stopping the firm's payment, and the term of bankruptcy, the legal procedure.

[6] In France, the nuclear risk is insured by special regime, for 10 years and at the level of 600 million francs by the firm, then of 1500 million by the State and finally of 2520 million by the European Union.

[7] "Nuclear house generally competitive for a based-load production, but this advantage is less 'obvious' than in the study led in 1993" (DIGEC's study[1997, p. 49]).

[8] We are not interested in the property of this structure, in contexts where there is creation of Independent Power Producers on pools and where the firms' privatization question in the network industries settles in monopolistic competition.

[9] We shall speak indifferently about state of insolvency or bankruptcy. But we distinguish this situation of that arisen from the fear of an imminent bankruptcy, financial distress.

7

maximizes its project value according to its debt. It is a possible situation for EdF which is subjected to tax and which is confronted with a more and more competitive environment where occur bankruptcies of firms, for examples in 2001 bankruptcies of both PG&E and SocalEd utilities on the CalPX Californian pool and of the broker in energy, Enron.

The value of the levered project is equal to the value of the unlevered project (leverage), net of the difference between the advantage (tax effect) and the inconvenience (bankruptcy risk) generated by debt:

$$V(p_t, C) = V(p_t, 0) + TB(p_t, C) - BC(p_t, C) = D(p_t, C) + E(p_t, C), \qquad (3)$$

where $\quad V(p_t, C)$ is the net current value of the levered firm;

$V(p_t, 0) = p_t$ is the net current value of the unlevered firm;

$TB(p_t, C)$ is the current value of fiscal earning due to debt;

$BC(p_t, C)$ is the current value of bankruptcy cost;

$D(p_t, C)$ is the current value of debt;

$E(p_t, C)$ is the current value of equity;

C is the amount of the debt coupon.

According to the trade-off theory, the firm should not rather self-finance its project, so underlining the capital-intensive character of the nuclear investment. It has to take then simultaneously two decisions, one of financing and one of investment.

Funding hypotheses - We suppose, following Modigliani and Miller[1958], Miller[1977], Brennan and Schwartz[1978], Leland[1994], that:

- the production activities of the firm remain unchanged by its capital structure. This hypothesis pushes aside any asset substitution problem (cf. Leland[1994, p. 1216]): the stockholders cannot choose riskier activities to increase the equity value to the detriment of the debt value and so hijack this value in their profit to the detriment of the bondholders (Harris and Raviv[1991]). Indeed, taxes and bankruptcy costs influence the optimal capital structure, even in case of possible asset substitutions[10]. We find the conclusion of Mauer and Triantis[1994] according to which the financial flexibility has, for the main part, no effect on the firm's production policy;

- the decisions concerning this financing structure are not modifiable anymore when they are taken. The debt value does not modify in the time and is defined as exogenous. We thus put aside all the agency problems between actors[11]: there is no issue of supplementary

---

[10] This effect influence is comparable to that of the risk of the firm's activity σ on the equity value and the debt value.

[11] These agency problems are nevertheless taken into account by the revisited trade-off theory, extension of the trade-off theory.

8

debt (to the detriment of the current bondholders), nor additional debt reduction (these repurchases being made to the detriment of the current shareholders), even if there are no refunding costs;

- the firm gets into debt only at the investing time, i.e. that the capital structure is fixed on all the duration of project's exploitation. Modigliani-Miller[1958],Merton[1977],Black and Cox[1976] put that the hypothesis of time-independent debt[12]. Now an innovation of the capital markets, the bullet repayment, makes possible the refinancement which allows to spread out the debt liability over the life expectancy of the nuclear equipment. We suppose so the refinancement costs are zero. We assume a debt structure with time-independant payouts;

- the optimal debt is determined in the presence of:

  * the corporate tax (CT) τ, $1 < \tau < 0$, which is equal to $37,77\%$ for EdF;

  * the firm goes bankrupt for a cost price p_t

    $6\ p_B$, but the current bankruptcy situation is estimated for $p_t \equiv p_B$. The bankruptcy characteristics are the following ones. The current bankruptcy cost, which includes the cost of the legal intervention,is fixed to α, $1 < \alpha < 0$. The market value of the firm is then lower than the amount of the debt: $V(p_B) = (1-\alpha)\,p_B < D(p_B)$. Finally, during the bankruptcy, the firm's property is transferred from the shareholders to the bondholders, and the shareholders are freed from any commitment: their liability is limited and $E(p_B) = 0$;

- the unprotected debt is risky, i.e. that the obligations have no protective covenants. This hypothesis implies that the bankruptcy price p_B is endogenous. The increase of the firm's activity risk σ increases the equity value $E(p)$, while lowering debt value $D(p)$. The shareholders should rather increase this risk, to the detriment of the bondholders, so as to maximize the equity value of the project. Making it $\frac{\partial E}{\partial p_B}, _{p_B = 0}$ thus defines itself as it is the

By this possibility of capital's property transfer of the equipment, we notice that to get into debt means buying a put option on project equity[13]. By levering, the shareholders give up a part of the equity property to the bondholders. They will get back this property by paying off the debt. The equity value is thus equal to the value of this put option for the shareholders. *It is only having exercised its real option (by investing) that the firm possesses this financial option (by getting into debt).*

---

[12] The debt maturity is infinite.

[13] The equity function is thus convex: $\dfrac{\partial^2 E(p)}{\partial p^2} > 0.$

9

3 The project reversibility: the option premium value

The optimally levered firm, according to the trade-off theory, tries to determine the price from which it invests. The firm adopts a feedback method. It calculates first of all the optimal amount of debt which maximizes its project value. It determines then the price threshold of accepting its maximum valued equipment. It so estimates its financial option before its real option: the opportunity is the sum of multiple interacting options (cf. Brennan and Schwartz[1985]), because the addition of the put on equity increases the real call value.

The difference between the cost price threshold in uncertainty (on the pool) and that obtained by the NPV rule in certain environment (in monopoly) characterizes the reversibility of the levered project. This comparison is thus made between the rules of the theory of real options and of the NPV which must necessarily be particularized.

3.1 The monopoly case

The cost price movement which characterizes the cost-of-service regulated monopolistic sector, was defined by (1):

$$\frac{dp}{p_t} = \bar{\mu} dt.$$

We are going to particularize the NPV rule. The interest of such a rule is to serve as base of comparison between the investment thresholds determined in this certain scenario of monopoly and the uncertain scenario of the pool. We considered that in certain environment, the firm which invests knows the optimal debt value of its project, the value of fiscal earning which it realizes by getting into debt and the cost which would provoke the transfer of its asset. This NPV rule is particular because we integrate the biggest possible values of fiscal earning and bankruptcy cost. This rule being kept in the pool scenario, the premium option value is not questioned.

The value of the optimally levered project $V^*(p)$ for the current price -

| Contracted at the investment time, the optimal debt $D^*(p_t)$ is perfectly known. It is constant during the NPP's exploitation and gets interests paid in the form of immediate,optimal and constant coupon, C^*:

$$D^*(p_t) = \frac{C^*}{r}.$$

| The current value of the fiscal earning of the debt, which the firm realizes if it is not acquired by its bondholders, is equal to:

$$TB(p_t) = \tau \frac{C^*}{r}.$$

10

Indeed, the fiscal legislation foresees that the debt interest, i.e. the coupon, is deductible from taxes for every exploitation exercise.

| In the current time, the firm also knows itscurrent bankruptcy cost:

$$BC(p_t) = \alpha p_B^*,$$

where p_B^* is the bankruptcy price defined for the optimal debt level D^*.

| The current optimally levered project value is equal to that of the unlevered equipment, i.e. the current cost price, to which adds the net fiscal earning of the cost of bankruptcy. The firm maximizes its optimal present project value net of the initial cost of capital:

$$V^*(p_t) = \max E\left[\left(p_0 e^{\bar{\mu}t} + \tau\frac{C^*}{r} - \alpha p_B^* - K\right)e^{-rt}\right], \tag{4}$$

where $\bar{\mu} \, \mathbf{6} \, 0 \implies \lim_{t\to\infty} p_0 e^{\left(\bar{\mu}-r\right)t} = 0$, the condition of McDonald and Siegel[1986] is always verified.

Because the regulator incites the firm to fix the price to be equal to the long term marginal cost (i.e. $\bar{\mu} \, \mathbf{6} \, 0$), the firm decides on its investment and on its financing in $t^* = 0$ because $V^*(p_1) \, \mathbf{6} \, V^*(p_0)$. The optimal projetc value (4) writes again:

$$V^*(p_0) = \max_{\left\{\vphantom{p_0}\right.} p_0 + \tau\frac{C^*}{r} - \alpha p_B^* - K; 0 \right\}. \tag{5}$$

To find the optimal debt which maximizes the project value at the time of the investment decision $t^* = 0$, we determine by means of the boundary conditions the bankruptcy price. Once the optimal debt coupon was estimated, we shall deduct the investment rule.

The asset's transfer price as the investment threshold - The asset's transfer price p_B^* is defined as the price of which the firm goes bankrupt. It is estimated for the optimal level of debt known by the firm at the time of its investment, that is at the initial time $t^* = 0$. If the firm goes bankrupt in the initial period, its capital value is transferred to its bondholders:

$$K = V^*(p_B^*) = D^* = \frac{C^*}{r}.$$

The levered firm value (5) is also expressed as:

$$V^*(p_B^*) = D^* = K = p_B^* + \tau K - \alpha p_B^*$$
$$\Leftrightarrow p_B^* = \frac{1-\tau}{1-\alpha}K.$$

This asset's transfer price p_B^* allows to estimate the put on the firm's equity. It corresponds, in this NPV rule, to the cost price threshold of investment acceptance:

$$p_{VAN}^* = p_0^* = p_B^* = \frac{1-\tau}{1-\alpha}K. \tag{6}$$

11

The funding structure - The decision of accepting or refusing the project being taken in the initial date $t^* = 0$, the value of the optimally levered project is obtained for $p_0 > p_B$ from the expressions (5) and (6):

$$V^*(p_0) = p_0 + \tau \frac{C^*}{r} - \alpha \frac{1-\tau}{1-\alpha} K.$$

Called θ, the optimal proportion of the debt value in the firm value: $0 \leqslant \theta \leqslant 1$. We can write again the levered project value:

$$V^*(p_0) = p_0 + \tau \theta V^*(p_0) - \alpha \frac{1-\tau}{1-\alpha} K.$$

From it, we deduct the optimal proportion of debt in the project value:

$$\iff \theta = \frac{1}{\tau} \left[1 - \frac{p_0 - \alpha \frac{1-\tau}{1-\alpha} K}{V^*(p_0)} \right] < 1.$$

This proportion expresses the funding choice of the firm. If the investment is held, it is levered at a θ percentage.

The investment rule - The decision of accepting or refusing the project is taken at the current time $t^* = 0$: it results from a "now or never" strategy. We determined by (6) the optimal cost price which implements the investment:

$$p^*_{VAN} = \frac{1-\tau}{1-\alpha} K, \tag{7}$$

where $p^*_{VAN} > K \iff \tau < \alpha$.

The rule of investment is the following one:

$$\begin{cases} \text{if } p_0 \leqslant p^*_{VAN}, \text{ then the firm does not invest and loses itself investment opportunity.} \end{cases}$$

The value of the nuclear project depends on the investment and its financing rules:

$$V_0^* = V^*(p_0) = \begin{cases} 0 \text{ if } p_0 \leqslant p^*_{VAN} = \frac{1-\tau}{1-\alpha} K \\ \qquad \text{with } D^*(p_0) = 0; \\ p_0 - \frac{1-\tau}{1-\alpha} K \text{ if } p_0 > p^*_{VAN} \\ \qquad \text{with } D^*(p_0) = \theta. \end{cases}$$

Proposition 1 *In the presence of both imperfections of capital market that are the tax rate and the bankrupcy cost of the firm for the trade-off theory, the monopoly can be optimally levered.*

a. The firm invests if the initial cost price is superior to the price threshold. Its capital structure correspond then to the constant leverage θ.

b. This price threshold is more than it is for an unlevered firm K, if the bankruptcy cost is superior to the tax rate.

12

In the certain environment, the investment threshold of the levered firm is more than that of the unlevered project if $\alpha > \tau$: the asset's transfer cost is superior to the tax rate for this specific NPV rule.

The questioning of the Modigliani-Miller[1958] hypothesis of separation of the financing and investment decisions has a first effect, that to move the accepting boundary of an optimally levered project with regard to the initial capital K, that is the accepting threshold of the levered project.

3.2 The pool case

3.2.1 $V^*(p)$ for the current price

The firm produces on the pool, the organization retained to introduce the competition on the production sector. It bases its believes on a cost price evolution which follows the decline of the long term marginal cost due to competition, according to a geometrical brownian motion (2):

$$\frac{dp}{p_t} = \bar{\mu}dt + \sigma dz.$$

The maximisation program of the project value is obtained according to our hypotheses:

$$V^*(p_t) = \max_{p_t}\left\{(1 - rdt)V^*(p_t) + E[dV^*]; 0\right\}.$$

The firm has to resolve the ordinary differential equation (ODE):

$$rV^*(p_t) = \frac{E[dV^*]}{dt} = \frac{dV^*}{dt}.$$

a) The project value $V(p)$

Because of the property of séparabilité additive of the value in absence of opportunity of arbitrage (AOA)[14], it is equivalent to resolve the following ODE:

$$rf(p_t) + c = \frac{E[df]}{dt} = \frac{df}{dt},$$

for $f = TB(p_t)$, $f = BC(p_t)$ and $f = D(p_t)$, because $V(p_t) = p_t + TB(p_t) - BC(p_t) = E(p_t) + D(p_t)$ (3). The constant c always represents the current flow of the function f.

If the marks do not indicate more than the derivates, we can write again the ODE with the Itô's lemma:

$$\frac{1}{2}\sigma^2 p^2 f_{pp} + \bar{\mu}p f_p + rf + c = 0. \tag{8}$$

We know that the solution shape is: $f(p) = A_0 + A_1 p^\beta$.

---

[14]It is equivalent to suppose the AOA of market or the additive separability of value. Now the AOA hypothesis is necessary to estimate the options.

Let us derivate:

$$\begin{cases} f_p = A_1 \beta p^{\beta-1}; \\ f_{pp} = A_1 \beta(\beta - \end{cases}$$

By putting these values in the expression (8), we obtain the relation:

$$A_1 \beta(\beta - 1)\frac{1}{2}\sigma^2 p^\beta + A_1 \beta \bar{\mu} p^\beta + r\left(A_1 p^\beta + A_0\right) + c = 0.$$

We don't care the obvious solution $p = 0$:

$$A_1\left[\beta(\beta-1)\frac{1}{2}\sigma^2 + \beta\bar{\mu} + r\right] = 0 \text{ et } rA_0 + c = 0 \Longleftrightarrow A_0 = \frac{c}{r}.$$

If $A_1 \neq 0$, $\frac{1}{2}\sigma^2\beta^2 + \left(\bar{\mu} - \frac{1}{2}\sigma^2\right)\beta + r = 0$ for the two following roots:

$$\begin{cases} \beta_N = \frac{1}{2} - \frac{\bar{\mu}}{\sigma^2} - \sqrt{\left(\frac{\bar{\mu}}{\sigma^2} - \frac{1}{2}\right)^2 + \frac{2r}{\sigma^2}} < 0; \\ \\ \beta_P = \frac{1}{2} - \frac{\bar{\mu}}{\sigma^2} + \sqrt{\left(\frac{\bar{\mu}}{\sigma^2} - \frac{1}{2}\right)^2 + \frac{2r}{\sigma^2}} > 0. \end{cases}$$

This ODE solution is equal to:

$$f(p) = A_0 + A_1 p^{\beta_N} + A_2 p^{\beta_P}.$$

We calculate the project value (3) $V(p) = p + TB(p) - BC(p) = E(p) - D(p)$, by resuming the various functions f: $f = TB(p_t)$, $f = BC(p_t)$ and $f = D(p_t)$. To obtain a finite project value (cf. infra the boundary conditions of $f(.)$), we are going to hold only the negative root β_N. We notice[15] that the increase of σ involves that of β_N and that the increase of the trend $\bar{\mu}$ or the interest rate r decreases β_N.

a.1) The fiscal earning

---

[15] $\frac{\partial \beta_N}{\partial \bar{\mu}} = \dfrac{\left(-\frac{\bar{\mu}}{\sigma^2} + \frac{1}{2}\right) - \sqrt{\left(\frac{\bar{\mu}}{\sigma^2} - \frac{1}{2}\right)^2 + \frac{2r}{\sigma^2}}}{\sigma^2\sqrt{\left(\frac{\bar{\mu}}{\sigma^2} - \frac{1}{2}\right)^2 + \frac{2r}{\sigma^2}}} < 0$ - Proof: $\sqrt{\left(\frac{\bar{\mu}}{\sigma^2} - \frac{1}{2}\right)^2 + \frac{2r}{\sigma^2}} \overset{>}{} \left(-\frac{\bar{\mu}}{\sigma^2} + \frac{1}{2}\right)$;

$\frac{\partial \beta_N}{\partial r} = -\dfrac{1}{\sigma^2\sqrt{\left(\frac{\bar{\mu}}{\sigma^2} - \frac{1}{2}\right)^2 + \frac{2r}{\sigma^2}}} < 0.$

$\frac{\partial \beta_N}{\partial \sigma} > 0$ has the same sign than $\frac{\partial \beta_N}{\partial \Sigma}$: $\frac{\partial \Sigma}{\partial \sigma} = 2\sigma$ and $\frac{\partial \beta_N}{\partial \Sigma} = \dfrac{2\bar{\mu}\sqrt{\left(\frac{\bar{\mu}}{\sigma^2} - \frac{1}{2}\right)^2 + \frac{2r}{\sigma^2}} + 2\left[\bar{\mu}\left(\frac{\bar{\mu}}{\sigma^2} - \frac{1}{2}\right) + r\right]}{\sigma^3\sqrt{\left(\frac{\bar{\mu}}{\sigma^2} - \frac{1}{2}\right)^2 + \frac{2r}{\sigma^2}}} > 0$ -

Proof: $\frac{\partial \beta_N}{\partial \Sigma} = 0$ for the two roots $r = \bar{\mu} \lessgtr 0$ and $r = 0$. $\frac{\partial \beta_N}{\partial \Sigma} > 0$ without these roots, thus for $r > 0$.

14

We obtain the following expression of the fiscal earning, $TB(p) = f(p)$:

$$TB(p) = A_0 + A_1 p^{\beta_N} + A_2 p^{\beta_P}.$$

The fiscal earning is defined as a cost avoided by the nontaxation of the debt coupon, which grows with the cost price. At the bankruptcy price p_B, the project property is given up to the bondholders and the debt disappears. The firm does not benefit any more then from fiscal earning. On the other hand, for a price enough high, the firm is sure not to go bankrupt and to realize a fiscal earning calculated on the debt interests. We summarize these two continuation conditions concerning the financing in the following way:

- when $p \to \infty$, the fiscal earning depends on the present debt amount: $TB(p)$
$= A_0 + A_2 p^{\beta_P} \Longrightarrow \begin{cases} A_0 = \tau \dfrac{C}{r}, \\ A_2 = 0. \end{cases}$

- when $p = p_B$, there is no fiscal earning by definition:
$TB(p_B) = 0 = \tau \dfrac{C}{r} + A_1 p_B^{\beta_N} \Longrightarrow A_1 = -\tau \dfrac{C}{r} p_B^{-\beta_N}.$

The value of the fiscal earning in the current period is equal to:

$$TB(p) = \tau \frac{C}{r} \left[1 - \left(\frac{p}{p_B} \right)^{\beta_N} \right], \tag{9}$$

where we can interpret $\left(\frac{p}{p_B} \right)^{\beta}$ as a bayesian probability of bankruptcy, i.e. a belief on the bankruptcy risk when the cost price is p. Leland[1994, p. on 1219] resumed the probability explanation[16] of Black and Cox[1976, pp. 355-356]. He clarified $\left(\frac{p}{p_B} \right)^{\beta} = \int_0^\infty \Phi(t, p, p_B) e^{-\bar{\mu}t} dt$, where $\Phi(t, p, p_B)$ is the density of the first immediate passage from p to the limit cost price p_B .

a.2) The bankruptcy cost

The bankruptcy cost $BC(p) = f(p)$ is solution of the ODE (8):

$$BC(p) = A_0 + A_1 p^{\beta_N} + A_2 p^{\beta_P}.$$

The boundary conditions specify the proportion of the firm's value[17] spent by the bankrupt stake and the absence of this type of cost on a solvent firm:

- when $p \to \infty$, there is no risk of bankruptcy:
$BC(p) \to 0 = A_0 + A_2 p^{\beta_P} \Longrightarrow \begin{cases} A_0 = 0, \\ A_2 = 0. \end{cases}$

---

[16] The value of the risk-neutral firm p follows a gaussian ($\frac{dp}{p_t} = \bar{\mu}dt + \sigma dz$) distribution function $\Theta(t', p(t') \text{Á} t, p(t))$.

[17] We characterized the unlevered firm's value by the cost price.

15

- when $p = p_B$, by definition: $BC(p_B) = \alpha p_B = A_1 p_B^{\beta_N} \implies A_1 = \alpha p_B^{1-\beta_N}$.

We obtain the bankruptcy cost of the firm:

$$BC(p) = \alpha p_B \left(\frac{p}{p_B} \right)^{\beta_N}.$$

We deduct the project value when the levered firm:

$$V(p) = p + \tau \frac{C}{r} \left[1 - \left(\frac{p}{p_B} \right)^{\beta_N} \right] - \alpha p_B \left(\frac{p}{p_B} \right)^{\beta_N}, \tag{10}$$

which depends on the coupon and on the leverage. If $\alpha > 0$ or if $\tau > 0$, then it grows with r and the trend $\bar{\mu}$, and when the risk of the activity σ decreases.

a.3) The debt

La valeur du projet est aussi égale à celle de la dette et des fonds propres (3). La valeur de la dette est solution de l'ODE (8):

$$D(p) = A_0 + A_1 p^{\beta_N} + A_2 p^{\beta_P}.$$

La dette est perpétuelle et constante pendant toute la durée d'exploitation du projet. Quand l'entreprise fait faillite, les créanciers récupèrent la valeur de son capital nette du coût de faillite. Les conditions aux bornes sont les suivantes:

- quand $p \to \infty$, $D(p) \to \frac{C}{r} = A_0 + A_2 p^{\beta_P} \implies \begin{cases} A_0 = \frac{C}{r}, \\ A_2 = 0. \end{cases}$

- quand $p = p_B$, la valeur de l'entreprise lors de la faillite est nette du coût de faillite:
$$D(p_B) = (1-\alpha)p_B = \frac{C}{r} + A_1 p_B^{\beta_N} \implies A_1 = \left[(1-\alpha)p_B - \frac{C}{r} \right] p_B^{-\beta_N}.$$

La valeur de la dette est égale à:

$$D(p) = \frac{C}{r} + \left[(1-\alpha)p_B - \frac{C}{r} \right] \left(\frac{p}{p_B} \right)^{\beta_N}. \tag{11}$$

a.4) The equity

The debt value deducted from the project value gives the equity value:

$$E(p) = V(p) - D(p) = p - (1-\tau) \frac{C}{r} - \left[p_B - (1-\tau) \frac{C}{r} \right] \left(\frac{p}{p_B} \right)^{\beta_N}.$$

16

b) Endogenous value of p_B: the no protected debt

By getting into debt, the firm bought a put option on equity. The representative curve slope of this option allows us to express the continuation condition between the project value and the debt value. Green [1984] showed that it depends on the value of the firm's transfer price. It is convex for a endougenously definite price, what corresponds to a risky bond bought on the financial market. The smooth-pasting condition is expressed for $p \equiv p_B$.

We know, by hypothesis, that the behavior of the stockholders leads them to maximize their equity value. The smooth-pasting condition is so expressed for $p = p_B$: $\dfrac{dE}{dp}\Big|_{p=p_B} = 1 + \beta_N \left[(1-\tau)\dfrac{C}{r}\dfrac{1}{p_B} - 1 \right] = 0,$

and allows to determine the endogenous value of bankruptcy p_B such as:

$$\Longrightarrow p_B = (1-\tau)\frac{C}{r}\left(\frac{\beta_N}{\beta_N - 1}\right). \tag{12}$$

We can show that the endogenous value of the cost price p_B and the value of the insolvent project $V(p_B) = (1-\alpha)\,p_B$:

- are proportional to the coupon C;

- are independent from the current cost price p;

- decrease when the CT rate τ increases;

- p_B is independent from the bankruptcy cost α, but $V(p_B)$ decreases when the bankruptcy cost α increases;

- decrease when the risk-free interest rate r increases;

- increase when the risk σ decreases;

- increase with the trend $\overline{\mu}$.

By parametrizing the following expressions:

$$\begin{cases} m = \dfrac{\left[\dfrac{(1}{N}\right]^{-\beta}}{}^{\dfrac{(\beta_N)\beta_N)r}{}}; \\ h = \left(1 - \beta_N - \dfrac{\alpha(1-\tau)\beta_N}{\tau}\right)m > 0; \\ g = [1 - \beta_N + (1-\alpha)(1-\tau)\beta_N]\,m > 0, \end{cases}$$

and by replacing the bankruptcy value (12), we can write again the project value:

$$V(p) = p + \tau\frac{C}{r}\left\{1 - h\left(\frac{C}{p}\right)^{-\beta_N}\right\}, \tag{13}$$

17

and the debt value:

$$D(p) = \frac{C}{r}\left\{1 - \left(\frac{C}{p}\right)^{-\beta_N} g\right\}.$$

(14)

We demonstrate the effects of the various parameters on the debt value:

- the increase of the risk-free interest rate has two effects on the debt value:

  * it increases the debt cost and decreases $D(p)$;
  * it increases the fiscal earning and $D(p)$.

  The first effect takes it on the second: the increase of the interest rate r lowers the debt value $D(p)$, while it is the opposite for an optimal leverage $L^* = \frac{D*(p)}{V*(p)}$;

- the debt value $D(p)$ increases with the coupon C, the cost price trend $\bar{\mu}$ and with the decline of the risk of the firm's activity σ.

We clarify these effects by taking into account the risk of bankruptcy:

- the increase of the bankrupcy cost α decreases $D(p)$;

- the increase of the CT rate τ lowers the endogenous bankruptcy price p_B, what implies the increase of $D(p)$;

- if $(\alpha; \tau) > (0; 0)$ and $p \to p_B$,

  * the increase of $D(p)$ is involved by that of the volatility σ or the interest rate r (the second effect takes it on the first one near the bankruptcy);
  * the decline of $D(p)$ is due to the increase of the coupon C (there is a maximal debt value) or to that of the cost price trend $\bar{\mu}$.

The optimal coupon, lower than the maximal coupon[18] ($C^* < C_{max}$), maximize the project value $V(p)$ (13): $\frac{\partial V(p)}{\partial C} = \frac{\tau}{r}\left\{1 - h\left(\frac{C}{p}\right)^{-\beta_N}\right\} + \frac{\tau}{r}\left\{\beta_N h\left(\frac{C}{p}\right)^{-\beta_N}\right\} = 0$

$$\Longleftrightarrow C^* = p\left[(1 - \beta_N) h\right]^{\frac{1}{\beta_N}}.$$

(15)

We estimate the bankruptcy price (12) for the optimal funding structure:

$$p_B^* = p\frac{(1-\tau)}{r}\left(\frac{\beta_N \beta_N}{-1}\right)\left[(1 - \beta_N) h\right]^{\frac{1}{\beta_N}} = p\left[(1 - \beta_N) - \frac{\alpha(1-\tau)\beta_N}{\tau}\right]^{\frac{1}{\beta_N}}$$

$$\Longleftrightarrow p_B^* = p\left(\frac{h}{m}\right)^{\frac{1}{\beta_N}}.$$

(16)

---

[18] $C_{max}/\frac{\partial D(p)}{\partial C} = \frac{1}{r} - \frac{(1-\beta_N)}{r}\left(\frac{C}{p}\right)^{-\beta_N} g = 0 \Longrightarrow C_{max} = p\left[(1 - \beta_N) g\right]^{\frac{1}{\beta_N}}.$
If $C^* = p\left[(1 - \beta_N) h\right]^{\frac{1}{\beta_N}} < C_{max} = p\left[(1 - \beta_N) g\right]^{\frac{1}{\beta_N}}$
$\Longrightarrow h > g$, still verified because $\beta_N < 0$.

18

3.2.2 The optimal funding structure -

The optimal funding decision defines the optimal part between equity and loan, that is the optimal leverage.

The optimal debt (13) is endogenously determined:

$$D^*(p) = \frac{C^*}{r} \left\{ 1 - \left(\frac{C^*}{p}\right)^{-\beta_N} g \right\}.$$

We substitute the bankruptcy price and the coupon by their optimal value (16) and (15):

$$D^*(p) = p \left[(1 - \beta_N) h\right]^{\frac{1}{\beta_N}} \frac{1}{r} \left\{ 1 - \frac{1}{(1 - \beta_N)} \frac{g}{h} \right\}.$$

To obtain the optimal equity value or optimal leverage, we have to calculate the optimal project value (10):

$$V^*(p) = p + \tau \frac{C^*}{r} \left\{ 1 - h \left(\frac{C^*}{p}\right)^{-\beta_N} \right\}$$

$$\iff V^*(p) = p \left\{ 1 + \frac{\tau}{r} \left[(1 - \beta_N) h\right]^{\frac{1}{\beta_N}} \frac{\beta_N}{\beta_N - 1} \right\}. \tag{17}$$

The optimal equity value is thus equal to:

$$E^*(p) = V^*(p) - D^*(p) = p \left\{ 1 - \frac{1}{r} \left[(1 - \beta_N) h\right]^{\frac{1}{\beta_N}} \left[1 - \frac{1}{(1 - \beta_N)} \left(\frac{g}{h} - \tau \beta_N\right)\right] \right\}.$$

We express the properties of this optimal leverage:

$$L^* = \frac{D^*(p)}{V^*(p)} = \frac{(1 - \beta_N) h - g}{r \left[(1 - \beta_N) h\right]^{\frac{\beta_N - 1}{\beta_N}} - \tau \beta_N h} < 1.$$

It does not depend on the current cost price p.

We find results corresponding to those waited:

- the firm having a few or very risky (according to the value of σ) activity pays, at the optimum, a higher coupon than a firm of which activity risk is average;

- at the optimum, the leverage of a riskier firm is always lower than that of a less risky firm;

- the optimal leverage is lower than 100%, even when the bankruptcy cost is zero. This result confirms that obtained by Brennan and Schwartz[1978].

The empirical results established by Leland[1994] give a optimal leverage value:

19

- if the firm's activity is few or averagely risky and if the bankruptcy cost α is average, $L^* \in [75\%; 95\%]$;

- if the activity risk and the bankruptcy cost are high and:

  * if the CT rate is of 35%, $L^* \in [50\%; 60\%]$;

  * if the CT rate is of 15%, $L^* \in [5\%; 25\%]$.

3.2.3 The optimal decision of investment -

The firm invests when the value of its investment option is equal to the project value (continuation condition):

$$V^*(p^*) = p^* - K + \frac{\tau}{r}C^* - \alpha p_B^*. \tag{18}$$

a) The investment rule

The expression of the optimally financed investment opportunity $V^*(p^*)$ is also given by (17). We write again (18):

$$p^* \left\{ 1 + \frac{\tau}{r}\left[(1-\beta_N)h\right]^{\frac{1}{\beta_N}} \frac{\beta_N\beta_N}{-1} \right\} = p^* \left\{ 1 + \frac{1}{r}\left[(1-\beta_N)h\right]^{\frac{1}{\beta_N}} \left[\tau - \alpha(1-\tau)\left(\frac{\beta_N\beta_N}{-1}\right)\right] \right\} - K$$

$$\Longrightarrow p^* = \left\{ \frac{(1-\tau)\beta_N}{\left(1 - \beta_N - \frac{\alpha(1-\tau)\beta_N}{\tau}\right)^{\frac{1}{\beta_N}}\left[-\tau - \alpha\beta_N(1-\tau)\right]} \right\} K,$$

the threshold price of accepting investment, which is positive if:

$$0 > \beta_N > -\frac{\tau}{\alpha(1-\tau)}.$$

The levered project value depends on the investment rule:

$$V_t^* = \begin{cases} 0 \text{ if } p_t \mathbf{6} \, p^* \\ \qquad\qquad\qquad \text{with } D^*(p_t) = 0; \\ p\left\{1 + \frac{\tau}{r}\left[(1-\beta_N)h\right]^{\frac{1}{\beta}}\frac{\beta_N}{\beta_N-1}\right\} \text{ if } p_0 > p^* \\ \text{with } D^*(p) = p\left[(1-\beta_N)h\right]^{\frac{1}{\beta}}\frac{t}{r}\left\{1 - \frac{1}{(1-\beta_N)}\frac{g}{h}\right\}. \end{cases}$$

b) The option premium Δ

The option premium value is equal to the difference between p^* and p_{VAN}^*:

$$\Delta = \frac{(1-\tau)\beta_N}{\left(1 - \beta_N - \frac{\alpha(1-\tau)\beta_N}{\tau}\right)^{\frac{1}{\beta_N}}\left[-\tau - \alpha\beta_N(1-\tau)\right]} - \frac{1-\tau}{1-\alpha}.$$

We find the same conclusions as in the previous case: the value of the multiple of the value of wait, which is that of the flexibility left with the firm in the timing of its investment and its financing, can be lower than the unity.

The price threshold of investment:

20

- increase with the bankruptcy cost;

- decrease when the CT rate increases:

  * if $\left(\alpha^2 + \alpha - 1\right)\tau^2 - \alpha\left(1 + \alpha\right)\tau + \alpha^2 < 0$;
  * or if $\left(\alpha^2 + \alpha - 1\right)\tau^2 - \alpha\left(1 + \alpha\right)\tau + \alpha^2 > 0$ et $0 > \beta_N > \frac{(1-\alpha)\tau^2 + \alpha}{\tau(\alpha^2 + \alpha - 1)^2 - \alpha(1+\alpha)\tau + \alpha^2}$;

  and increase with the CT rate:

  * if $\left(\alpha^2 + \alpha - 1\right)\tau^2 - \alpha\left(1 + \alpha\right)\tau + \alpha^2 > 0$ et $\beta_N < \frac{(1-\alpha)\tau^2 + \alpha}{\tau(\alpha^2 + \alpha - 1)^2 - \alpha(1+\alpha)\tau + \alpha^2} < 0$;

- decrease, as the flexibility value of the firm, when β_N increases, i.e. when the risk σ increases and when the risk-free interest rate r or the trend $\bar{\mu}$ decrease.

Proposition 2 *On the pool, the competitive firm invests beyond a price threshold p^* which can be lower, more equal or superior to that determined in the certain case p^*_{VAN}.*

The option premium Δ can be negative. The firm invests in a negative NPV project to know, right now, the tax rate and the transfer cost of its asset and to define so its funding structure.

Corollary 3 *The option premium Δ decreases when the cost price trend $\bar{\mu}$ and the interest rate r decrease. It also lowers for an increase of the risk of the firm's activity σ. The investment threshold p^* gets closer to that determined in the certain case p^*_{VAN}.*

If the firm can finance its project by a loan, we notice that its investment threshold p^*_{VAN} (defined by (7)) is fixed according to the values of the rate of CT τ and the cost of bankruptcy α. In uncertainty, the threshold of release of the investment p^* still depends on these two exogeneous variables in the model, but also parameters of the evolution of the cost price rate. The fact that the investment threshold p^* is variable with regard to the threshold determined by the NPV rule and depends on values of the various parameters, characterizes the reversibility of the investment opportunity. An option premium value can be negative, positive or zero because the irreversible project becomes reversible (cf. Abel and al.[1996], Dixit and Pindyck[1998]) when its financing is taken into account.

4 Conclusions and extensions

4.1 Conclusions

In this model, we characterized the deregulation by the uncertainty of the variation rate of the cost price. Furthermore, in this environment, the firm resorts to debt to finance its investment.

21

The firm has the opportunity to invest in a nuclear equipment which is comparable to a real option. This investment is a totally irreversible option to defer if it is unlevered, i.e. in the hypothesis of Modigliani-Miller[1958] of separation of investment and financing decisions. In finance, this hypothesis is questioned by the trade-off theory which puts in motto the interaction between the investment and financing decisions. We use this theory to estimate the investment opportunity as well as its debt, because the firm establishes the corporate financing structure to isolate risk.

We notice that the optimal debt increases the project value with regard to the unlevered project value, because the debt interests are deductible from taxes and because the equipment in question can be given up. The real option is not anymore an option of defer, defined by the choice terms in self-financing: wait or invest. The consideration of the optimal leverage adds a financial put option to the real call option which increases the opportunity value: there are multiple interacting options. This put estimates the flexibility of the settled project, i.e. that it defines the capacity of the firm's reaction in front of uncertain events. It is exercised for a transfer price of the asset which is the endougenously definite bankruptcy price p_B.

We show that the introduction of the optimal debt modifies the option premium value: it is not still positive anymore, value which defines a totally irreversible opportunity. An optimally levered project is partially reversible. The possibility that the option premium value is negative reveals that the levered investment opportunity is partially reversible. Indeed, the loan implies the existence of a bankruptcy, a possibility of the asset's transfer to the bondholders.

The premium option value Δ in the trade-off theory:

Firm	Unlevered	Optimally levered
Value Δ	$\Delta > 1$	$\Delta = ?$
Capital	irreversible	partially reversible

The firm waits before investing until its project value compensates for the wait value of supplementary information about the future market conditions. This information also concerns its financing conditions.

When the option premium value is negative, it means that the firm invests for negative NPV project. Roberts and Weitzman[1981] interpreted this firm's anticipation as the fact that it tries to hold a certain information in the future. This information concerns its financing characteristics: the tax rate and the transfer cost of the asset, both imperfections of the capital markets held by the trade-off theory which impose upon the firm. It determines its bankruptcy price and the put value on the firm's equity. The nuclear opportunity consists of two options. The real option is exercised when the project is agreed and led the firm to buy an put on its equity by contracting its debt. Once the estimated put, we deduct the investment rule. We prove that the debt can lead the firm to anticipate the criterion of NPV.

22

4.1.1 Extensions

The dividends payment to the shareholders - We put until now the hypothesis of Brennan and Schwartz[1978] that there is no net cash flow's exit, notably because dividends are financed by issuing new actions in an amount equal to that of any supplementary dividend distribution (hypothesis verified on a perfect capital market). But these cash flow exits exist in imperfect market of the capital: they include dividends paid to the bondholders, and/or interest payments after tax, that remain lower than the financing by equity. Firms adopt a policy of fixed dividend. By defining a fixed target payout ratio, they assure the dividend constancy, or at least its gradual evolution.

If we admit (following Leland [on 1994]) that the net cash flows exit is from a proportional value to the asset value, the cash flows evolution (2) becomes equal to:

$$\frac{dp}{p_t} = \left(\bar{\mu} - d\right) dt + \sigma dz,$$

where d corresponds to dividends paid to the bondholders, this proportion depending on the debt coupon C.

If $d > 0$ (the dividend flow can be fixed with regard to the constant coupon), $d = (1-\tau)C + c$, c fixed), the optimal leverage decreases.

The existence of potential agency problems - The current debt rate fluctuates around the target return rate. This variation takes the shape, in the absence of negiotiation cost, of continuous adaptations of the debt coupon C so as to maximize the firm's value when the cost price varies.The debt coupon is not time-independant anymore. No repurchase (or issue) of debt being wished by the shareholders (or the bondholders), its adjustment is made by debt renegotiation: it takes into account agency costs.

References

[1] Abel A.B.[1983], Optimal investment under uncertainty, *American economic review* 73, 228-233

[2] Abel A.B. and Eberly J.C.[1997], An exact solution for investment and value of a firm facing uncertainty, adjustment costs, and irreversibility, *Journal of economic dynamics and control* 21, 831-852

[3] Abel A.B. and Eberly J.C.[1996], Optimal investment with costly reversibility, *Review of economic studies* 63 n°4, 581-593

[4] Abel A.B., Dixit A.K., Eberly J.C. and Pindyck R.S.[1996], Options, the value of capital, and investment, *Quarterly journal of economics* CXI n°3, 753-777

23

[5] Akerlof G.[1970], The market of lemons: quality uncertainty and the market mechanism, *Quarterly journal of economics* 89, 488-500

[6] Black F. and Cox J.[1976], Valuing corporate securities: somme effects of bond indenture provisions, *Journal of finance* 31, 351-367

[7] Brennan M. and Schwartz E.[1985], Evaluating natural resource investments, *Journal of business* 58 n°2, 135-157

[8] Brennan M. and Schwartz E.[1978], Corporate income taxes, valuation, and the problem of optimal capital structure, *Journal of business* 51 n°1, 103-114

[9] Caballero R.J.[1991], On the sign of the investment-uncertainty relationship, *American economic review* 81 n°1, 279-288

[10] DIGEC[1997], Les "coûts de référence" en production électrique, sous la direction de Batail J., Ministère de l'économie, des finances et de l'industrie, Secrétariat d'Etat à l'industrie, DGMP

[11] Dixit A.K.[1995], Irreversible investment with uncertainty and scale economies, *Journal of economic dynamics and control* 19, 327-350

[12] Dixit A.K. and Pindyck R.S.[1998], Expandibility, reversibility, and optimal capacity choice, NBER working paper n°6373

[13] Dixit A.K. and Pindyck R.S.[1994], Investment under uncertainty, Princeton university press, Princeton

[14] Green R.[1984], Investment incentives, debt and warrants, *Journal of financial economics* 13, 115-136

[15] Harris M. and Raviv A.[1991], The theory of capital structure, *The journal of finance* XLVI n°1, 297-355

[16] Leland H.E.[1994], Corporate debt value, bond covenants, and optimal capital structure, *The journal of finance* XLIX n°4, 1213-1252

[17] Lescoeur B. and Penz P.[1999], La problématique du financement des investissements électronucléaires, *Revue d'économie financière* 51, Le financement des infrastructures, 167-182

[18] Mauer D.C. and Triantis A.J.[1994], Interactions of corporate financing and investment decisions: a dynamic framework, *The journal of finance* XLIX n°4, 1253-1277

[19] McDonald R. and Siegel D.[1986], The value of waiting to invest, *The quarterly journal of economics* CI n°3, 707-727

24

[20] Merton R.C.[1977], On pricing of contingent claims and the Modigliani-Miller theorem, *Journal of financial economics* 5, 241-249

[21] Modigliani F. and Miller M.H.[1958], The cost of capital, corporation finance and the theory of investment, *American economic journal* 48 n°3, 261-297

[22] Myers S.C.[1977], Determinants of corporate borrowing, *Journal of financial economics* 5, 147-175

[23] Pindyck R.S.[1988], Irreversible investment, capacity choice and the value of the firm, *American economic review* 78 n°5, 969-985

[24] Roberts K. and Weitzman M.[1981], Funding criteria for research, development and exploration projects, *Econometrica* 49 n°5, 1261-1288

[25] Smith W.T.[1994], Investment, uncertainty and price stabilization schemes, *Journal of economic dynamics and control* 18, 561-579

[26] Trigeorgis L.[1996], Real options, The MIT press, Cambridge, Massachusetts

25

Real options theory and the broadcasting industry – a conceptual outline for potential application areas

Dipl.-Kfm. Marcus Dimpfel
mcm institute
University of St. Gallen
Blumenbergplatz 9
9000 St. Gallen
Tel.:   0041 (0)71 224 3084
Fax.:   0041 (0)71 224 3078
E-Mail: Marcus.Dimpfel@unisg.ch

Dipl.-Math. René Algesheimer
Center for Market Orientied Product and Production Management (CMPP)
Johannes-Gutenberg Universität Mainz
50099 Mainz
Tel.:   0049 (0)6131 392 2079
Fax.:   0049 (0)6131 392 3727
E-Mail: algeshei@mail.uni-mainz.de

1 Introduction

In the recent past, the research area of real options theory (ROT), in which researchers and practioners together are trying to seize and measure the construct of managerial flexibility, has gained increased importance (Stark (2000), p. 313). Although already known for a notable time in finance (Kester (1984); McDonald/Siegel (1986)), ROT roused the attention of more and more researchers from other business disciplines such as strategy (Amram/Kulatilaka (1999a); Amram/Kulatilaka (1999b)) or marketing (Hommel/Ludwig (1999); Kühn/Fuhrer (2001)). In addition to analysing the impact of ROT on a functional level, research also focuses on the identification of potential application areas on an industry level. In this context on the one side it is argued that the application of ROT is especially valuable for those industries, in which investments are characterised by a high degree of uncertainty and irreversibility (Meise, 1998, pp. 6). On the other side it is stated that the broadcasting industry and especially its television segment can be interpreted as such a business environment and that the digital convergence's impact even increases the complexity and dynamics faced by the relevant decision makers (Bughin, 2001, p. 64). As a consequence it is assumed that the application potential for ROT in the business context of the broadcasting industry is especially high.

The objective of our conceptual paper originates from this assumption as we want to analyse if and in what areas the application of ROT on the broadcasting industry is indeed advisable. Our paper's further proceeding is consistent with this objective and divided into three parts. In chapter two we will outline ROT as the theoretical foundation of our paper with respect to our specific concern. In chapter three we will analyse the research context of the broadcasting industry and thereby answer our research objective as stated before. Chapter four closes our analysis by offering the paper's conclusion and a discussion concerning further research.

2 Theoretical foundation: real options theory

In this chapter we will give an overview of our understanding and classification of real options and subsequently substantiate uncertainty and irreversibility as the abstract determinants of the relevance of action flexibility.

1

2.1 Classification of real options

Real options refer to different forms of action flexibility (Kulatilaka, 1995b, pp. 99-104), whereat we define the latter as "(…) the ability to change or react with little penalty in time, effort, cost or performance" (Upton, 1994, p. 73). In this paper we categorise the different forms of action flexibility as learn options, expansion options, assurance options and growth options.

Learn options allow a firm to tie its resource allocation to the solution of project-related risk and are associated to a point of time before the actual investment takes place, respectively is completed. Lern options can be subdivided into options to wait and options to stage investments. Options to wait allow a company to postpone the actual investment decision as the potential investment opportunity sustains for a certain period. During this time period new information can arise, reduce the project-related uncertainty and therefore change the economic feasibility of an investment. Waiting options result from the timely acquisition of licences, patents or other rights, which exclude competitors temporary or permanently from the focal investment opportunity. Options to wait can be understood as call options. Options to stage investments refer to the fact that most investments don't necessarily have to be financed by a single up-front outlay, but can be financed by a sequence of smaller amounts. In this sense companies have the possibility to refrain the further investments, if the costs for the next investment stage surmount the value of the continuing project. As a consequence options to stage investments can be interpreted as compound options (options on options). In this context each partial amount corresponds to the exercise price for acquiring the sequential option.

Expansion options and assurance options refer to forms of action flexibility, that exist during and after the actual investment phase and are related to an existing project. Expansion options empower a company to increase its economic activity, i.e. to expand production or distribution, depending on a positive development of the relevant economic parameter. Expansion options can be viewed as call options.

Contrarily, assurance options permit management to react to a negative development of the economic parameter with a reduction or modification (switching options) of its economic activity. Concerning the reduction, depending on its magnitude we distinguish between options to contract, options to shut down

2

and restart as well as options to abandon. The mentioned options can be interpreted as put options. Within the switching options, we further differentiate between options that allow a switch of the input factor and those options that enable management to produce alternative products.

Growth options refer to a time frame after the investment phase and refer to qualitatively innovative products. Growth options are of enormous strategic importance as a project might not appear profitable as a stand alone project, but may enable profitable future investments. The main value of growth options is therefore not related to the project's own cash flows, but to those of the potential future projects. Growth options can be viewed as call (compound) options.

2.2 Drivers of the relevance of real options

In our paper the relevance of action flexibility, resp. real options is interpreted as management's ex ante perception of the ratio between the expanded net present value (NPV), respectively the NPV including the consideration of action flexibility and the traditional NPV, respectively the NPV without the consideration of action flexibility. In this sense the relevance of action flexibility is high if the relevant decision makers estimate the investment to have a high option value component and therefore think that the disregard of the inherent action flexibility could lead to grave misinvestments.

The starting point of our analysis concerning the relevance of action flexibility is the awareness that an investment decision is trivial only in two cases (Meise 1998, pp. 6-11). Firstly, it is if the decision maker is perfectly informed with respect to all of the relevant data, because in this case the decision can be reduced to the simple calculation of the best alternative. Supplementary changes are not necessary. Secondly, the investment decision is trouble-free if the focal decision is completely reversible, that means if the decision can be revised without any costs. In those two scenarios action flexibility is of no specific use and could be neglected. The reverse means that the relevance of action flexibility for a decision context is determined by the combination of the degree of uncertainty and irreversibility. In the proceeding we therefore substantiate those abstract determinants of the relevance of action flexibility.

3

2.2.1 Uncertainty

With respect to the underlying reasons for uncertainty one can distinguish between complexity and dynamics. Complexity is point of time related and refers to the fact that the large number of relevant variables can assume different states and that their interdependencies change on the basis of those individual state configurations (Duncan, 1972, pp. 313-327). Dynamics results from the instance, that the states of the variables and the latters' interrelations are instabile over time. It can therefore be interpreted as complexity over time. The dynamics of a specific context is mainly influenced by the frequency, the intensity and the irregularity of the variable changes (Kieser/Kubicek, 1983, p. 319).

If a company wants to analyse the parameter value of a specific investment context's complexity and dynamics as a whole, it first of all has to identify the different textual sources of uncertainty. In this paper we consider the uncertainties concerning the consumers' preferences and needs, the actions of competitors and the technological development as the most important textual sources (Micalizzi/Trigeorgis, 1999, pp. 2-5).

Uncertainties with respect to the consumers are certainly most important as it is the primary goal of every company to interact with and sell their goods to their customers. In this context it has to be mentioned that depending on the considered goods, companies as well as consumers have to some extent measures, like e.g. screening or signaling activities, to reduce uncertainty. Within screening (Stiglitz, 1974, pp. 28-44), in which the less informed exchange party takes the initiative, one can distinguish between two versions. The activity of examination refers to a detailed analysis of the relevant variables through the less informed party, although it has to be mentioned that not all goods are equally accessible for such an inquiry. The activity of self selection aims on getting the better informed party to classify itself and in this context reveal the desired information. Signaling (Spence, 1973, p. 357) also refers to information transfers from the better to the less informed party, but this time the initiative stems from the better informed exchange party. In the consequence the amount of consumer related uncertainty is highest in those industries, where the consumer preferences and needs are statically very heterogenous and dynamically very volatil and where the goods produced give companies (and consumers) none or only limited room for reducing the uncertainties through the measures of screening and signaling.

4

Uncertainties concerning the competitors result from the instance that strategic or tactical moves like e.g. aggressive price competition, introduction of innovative products or merger's and acquisisition can have a huge effect on the market demand for a company's product, especially in oligopolistic industries.

The uncertainties concerning the further development of the existing and the occurrence of new technologies, is closely related to the competitive uncertainty as the combination of different technological factors can lead to a loss in competitive advantage. In this context is has to be mentioned that especially the process of technological innovation is per definitionem uncertain and that therefore one challenge for companies refers to the optimal timing of investment decisions in the innovation process.

2.2.2 Irreversibility

The degree of an investment project's irreversibility is mainly due to the amount and character of a company's fixed costs and therefore determined by two forces, its operating risk and the degree of the reselling potential of the involved resources.

The operating risk refers to this part of the variability of a company's profit that results from its operating structure and is mainly driven by a company's ratio of fixed and variable costs (Micalizzi/Trigeorgis, 1999, p. 2). In this sense it is widely accepted that a firm structure with a prevalence of fixed costs is ceteribus paribus rigid and difficult to modify if the relevant economic conditions change. In this context, the strength that the volatility of sales has for a company's operating results is referred to as its operating leverage.

The second force that drives an investment's irreversibility is the degree up to which the invested resources can be sold efficiently. In this context *Pyndick* (Pyndick, 1991, pp. 1110-1111) differentiates between the specificity of an investment, the efficiency of the second hand markets for the considered resources, political and legal regulations and the pressure from the public.

One of the most important reasons for the irreversibility of an investment is its specificity. In this context we understand specificity as the degree up to what resources are specially designed or located for a particular use or user (Masten, 1986, p. 494). The degree of specificity is measured by the amount of the quasi-rent, which labels the difference between the return of the invested ressources in

5

the considered disposition and the return in the best utilisation opportunity outside the focal disposition (Klein/Crawford/Alchian, p. 298). Furthermore we differentiate between different forms of specificity, namely industry specificity and firm specificity. An example for industry specific investments are steel producing facilities as they can only be used to produce steel. Although the equipment in general could be sold to other steel companies, especially in competitive industries the investment costs can mostly be viewed as irreversible as the value of the equipment will be about the same to all companies. As a consequence there is likely to be little gained by selling the resources. Firm specificity refers to those investments that are only of use for a specific company. Possible good examples can be found in human capital investments (knowledge transfer in trainings) or brand capital investments (brand or product specific investments). Those investments can't be reversed by selling them to competitors.

Another important driving force for the irreversibility of an investment is the degree of efficiency of the second hand markets for the involved resources. In this context especially the problem area of adverse selection is of great importance. In his famous acticle, Akerlof (Akerlof, 1970, pp. 488-500), building on the insights of information economics and principal-agent-theory explains the "lemon problem" in the market for used cars. The basis of his description is a market with heterogenous quality characteristics, which can't be identified by the buyer (principal) before the conclusion of the contract (quality uncertainty) and can't be influenced by anyone at the point of the contract's conclusion. As a consequence the buyer is not able to assess the quality of a good in a differentiated way and consequently only pays a price which amounts to the subjectively assumed average price. The sellers (agents) of goods with a higher than average quality will not accept this price and leave the market. As a consequence the average quality decreases and even more high quality sellers withdraw from the market. The last consequence of this phenomenon is the collapse of the market. On markets with such a "lemon problem" even non-specific goods, like office-equipment, cars or computers can only be sold well below their investment costs

As mentioned before other reasons for irreversibility can be found in regulations. In this sense capital control mechanisms can prohibit the selling of foreign direct investments or working law regulations can make human capital investments (temporary) irreversible. Furthermore, sometimes even the public opinion can

6

cause investments to be irreversible, e.g. in the case when a company wants to sell its pollution control equipment.

3 Real options theory and the broadcasting industry

In this chapter we first of all analyse if the application of ROT to the broadcasting industry in principal is advisable. Hence, we will examine the respective parameter values of uncertainty and irreversibility for this context. Afterwards we will deepen our analysis by highlighting potential application areas for ROT on the basis of the different segments of the broadcasting industry's value chain.

3.1 Relevance of real options in the broadcasting industry

Analogous to the preceding chapter we will differentiate our analysis regarding the relevance of real options in the broadcasting industry with respect to the parameter value of its uncertainty and its irreversibility

3.1.1 Uncertainty

In this section we will, in accordance to the above described differentiation, focus on the parameter values of uncertainties concerning the consumers' preferences and needs, the actions of competitors and the technological development. In this context we will firstly focus on uncertainties with respect to the traditional broadcasting business and will then highlight the digital convergence's impact on the overall amount of uncertainty.

Concerning the uncertainty regarding the consumer preferences and needs, the most significant characteristic of the broadcasting industry is that there is no direct relationship between the broadcasting company and the recipient (Zerdick et al., 1999, p. 55). Rather it is an indirect and anonymus, only through audience research institutions mediated relationsship. Therefore the focal uncertainties are per se very high, especially as there are no objective quality criteria for broadcasting products. As a consequence the affected companies can't utilise the measures of signaling and screening. Furthermore the audience is usually very volatile since it is composed of a wide range of socio-demographic segments, with widely heterogenous tastes. In addition the different preferences of the audience change rapidly and formats that used to attract viewers just a couple of month before, aren't working any more (i.e. reality formats). Although competition

7

concerning the audience traditionally is national or language specific, the different markets are already characterised by a strong competitition with continuously shifting market shares between the different governmental and private broadcasting stations. In addition the procurement market has always been global, wherefore the overall uncertainty regarding the competition, especially after the privatisation wave in Europe, is very high. Regarding the impact of technological development we can state that in the past particularly the distribution segment of the broadcasting industry's value chain has been strongly affected by major technological changes, i.e. cable TV or satellite-TV. However, especially in comparison to the before described uncertainties regarding the audience's preferences, the share that the technological uncertainty has in the overall complexity and dynamics was relatively low so far.

With the phenomenon of digital convergence becoming reality the relevance of technological uncertainty increases dramatically, thereby strongly influencing the other two categories of uncertainty. In this context especially the digitisation and the internet are expected to increase the average uncertainty in the broadcasting industry (Bughin, 2001, 69-71). Signal digitisation which has already been deployed in different markets, increases the capacity of channels significantly and enables new interactive services. With respect to the program diversity, digitisation presumably will increase the number of semi-generalist and thematic channels. As a consequence the audience will have a far wider variety of programs, which will lead to an even further fragmentation of the audiences. With respect to the internet, broadcasting executives have to take into consideration that a possible shift of leisure-time activity from the television to the internet in the long term puts pressure on the advertising attractiveness of broadcast media to advertisers. On the other hand the internet can serve as an additional distribution channel to syndicate a broadcaster's content to consumers.

Overall we can summarise that the level of uncertainty in the broadcasting industry is very high and will increase even further in the future.

3.1.2 Irreversibility

Companies in the broadcasting industry in general have a very high operating risk. This is due to the fact, that a very large portion of a broadcasting company's total costs refers to infrastructure and program purchasing, resp. production investments as well as marketing expenditures, all costs which are fixed in nature. Furthermore, the reselling potential of the resources invested is very low. In general, all of them are only of use in this particular industry and because of the intense competition can't be sold profitable to other stations. In this sense it is unlikely that existing equipment or content formats which haven't been profitable for one station will be profitable for another. In addition broadcasting content is usually sold in company specific bundles. Another reason for the low reselling potential of broadcasting products can be found in the before mentioned missing of objective quality criteria wherefore the price fixation for formats is difficult. Furthermore until its completion the costs of producing a company's own content are even more specific. The value of the "raw material" of the invested intangible resources, brought in by journalists and actors, can't be sold at all until the completion of a product and later on are also very firm-specific. Marketing expenditures are also classical examples for firm-specific investments as they can't be reversed by selling them to competitors. Moreover the prevailing share of human capital in combination with labour law regulations further enhances the degree of irreversibility.

Altogether we can resume that investments in the broadcasting industry are not only characterised by a high amount of uncertainty, but also by a high degree of irreversibility. As a consequence the relevance of action flexibility is especially high and the application of ROT onto this specific context seems advisable.

3.2 Real options in the value chain of the broadcasting industry

In this chapter we want to deepen the analysis concerning the relationship of ROT and the broadcasting industry by examining which real options are of relevance in the different segments of the broadcasting's value chain. Therefore we focus on the constituting segments of procurement, production, programming, distribution and end devices

3.2.1 Procurement

The procurement phase in the broadcasting value chain refers to the installation of the necessary infrastructure equipment, the procurement of (bundles of) ready for

9

use broadcasting formats and the acquisition of rights, necessary for beginning with an own production.

In this context especially learn options are of relevance. With respect to the tangible and intangible infrastructure they mainly refer to options to stage an investment as most of these investments don't necessarily have to be financed by a single-up front outlay, but can be splitted. Options to stage an investment are also very important with respect to the sourcing of telecasts. This is due to the fact that usually broadcasting companies don't have to buy all sequels at once, but can rather decide from season to season. In contrast options to wait most notably refer to rights for the production of own content, e.g. the possibility to postpone the start of the production of a movie or its sequel.

3.2.2 Production

Within the production phase, broadcasting companies (co-) produce content on the basis of the rights acquired in the procurement phase. In doing so the companies usually possess options to stage investments. In this sense the normal production process is divided into the pre-production, the actual production and the post-production process and broadcasting companies can decide whether to continue or not with the project at any point of time. With respect to telecasts produced by the broadcasting company, the production segment contains also assurance options as the company can (temporary) abandon the production of sequels due to a negative viewer's response. Furthermore some broadcasting formats contain expansion options as its production volume can easily be increased due to a surprisingly positive development of the audience's appeal (e.g. talkshows or late-night shows). The production segment can also hold growth options, which exist if the additional production volume is not related to the same, but to new formats. A possibly good example can be found in spin-offs of existing TV series, which are based on already established characters (e.g. "Cheers" and "Frasier").

3.2.3 Programming

In this segment of a broadcasting company's value chain the different sourced and (co-) produced formats are combined to create a company's program. In this context broadcasting companies have the opportunity to vary the time slot for a

10

specific format. Furthermore, if a broadcasting company owns more than one channel, it can move formats from one channel to the program of a more adequate channel. In addition the programming segment also possesses assurance options. In this sense companies can stop the transmission of expensive formats temporary, e.g. in the summer time, when the audience appeal in general is lower due to outdoor activities of the potential audience. Options to abandon the programming of an already acquired format prematurely are only relevant in those cases in which the programming space is very limited and therefore the opportunity costs are very high.

3.2.4 Distribution

The distribution as well as the successional segment of end devices are typically not part of a broadcasting company's core business, but have to be taken into account. The distribution refers to the transmission of a broadcasting company's program via alternative infrastructures. In this context the relevance of options mainly refers to expansion and switiching options. Expansion options refer to idle frequencies or bandwith that a broadcasting company can leverage if the designed program has a positive audience appeal and therefore should be made accessible to a wider range of viewers. Switching options in the distribution segment of the broadcasting business are related to alternative distribution platforms, e.g. terrestrial, cable, satellite or the internet. If a company possesses access to the different infrastructures, it can switch to the prevailing standard in the future.

3.2.5 End devices

The market for end devices is relevant in as much as it is closely related to the distribution of a broadcasting company's content and besides the available bandwith represents the bottle-neck concerning technologically advanced forms of broadcasting (e.g. interactive TV). As the future standards are not yet set, broadcasting company's somehow have to aquire switching options, so that they can switch to the prevailing standards. Furthermore, the market for end devices typically contains platform investment related growth options. In this sense broadcasting companies subsidise settop boxes which allow a two-way communication between the viewer and the broadcasting company and in the consequence pave the wave for potential future markets like i.e. t-commerce.

11

4 Conclusion

After sketching the paper's theoretical foundations we have first of all analysed the parameter values of the abstract drivers of the relevance of real options, uncertainty and irreversibility, in the broadcasting industry. We concluded that the focal environmental context is characterised by a high degree of uncertainty and irreversibility and therefore deepened the analysis concerning the relationship between real options theory and the broadcasting industry. In this context we have examined the different segments of the broadcasting industry's value chain. As a resume we can state that real options are important in each of the different segments and because of the impact of digital convergence will even get more important in the future. As a consequence we conclude that the consideration of real options in the strategic management of broadcasting companies can lead to a significant increase in profitability.

However, we have to mention that the postulated results of our conceptual paper tentatively remain theoretical in nature. Future analyses should therefore focus on the operationalisation of the relevant constructs and dimensions as well as on empirical testing. Furthermore, research should focus on potential pitfalls with the actual implementation of ROT in the broadcasting industry. In this context *Bughin* suggests the implementation of a real options process, which features the option inventory as the collection of all necessary information and the option heuristics which labels a process that evaluates the option value along the option period and decides on exercising options based on clear-cut and heuristic rules defined by the channel manager (Bughin, 2001, pp. 77).

12

Bibliography

Akerlof, G.A.: The market for lemons: qualitative uncertainty and the market mechanism, in: Quarterly Journal of Economics, Vol. 84 (1970), No. 3, pp. 488-500.

Amram, M./Kulatilaka, N.: Disciplined decisions. Aligning strategy with the financial markets, in: Harvard Business Review, Vol. 77 (1999a), No. 1, pp. 95-105.

Amram, M./Kulatilaka, N.: Real Options. Managing strategic investments in an uncertain world, Boston 1999b.

Bughin, J.: Managing Real Options in Broadcasting, in: Communications & Strategies, Vol. 41 (2001), No. 1, p. 63-78.

Duncan, R.B.: Characteristics of organizational environments and perceived environmental uncertainty, in: Administrative Science Quarterly, Vol. 17 (1992), No. 3, pp. 313-327.

Hommel, U./Ludwig, A.: Die Bewertung von Markteintrittsstrategien mit dem Realoptionenansatz, in: Zentes, J. (Hrsg.): Fallstudien zum internationalen Management, Wiesbaden 1999, pp. 535-543.

Kester, W.C.: Today's options for tomorrows growth, in: Harvard Business Review, Vol. 62 (1984), No. 2, pp. 153-160.

Kieser, A./Kubicek, H.: Organisation, 2nd edition, Berlin, 1983.

Klein, B./Crawford, R./Alchian, A.A.: Vertical Integration, Appropriable rents and the competitive contracting process, in: Journal of law and economics, Vol. 21 (1978), No. 2, pp. 297-326.

Kühn, R./Fuhrer, U.: Die Bedeutung von Realen Optionen für Marketingentscheidungen, in: Journal für Betriebswirtschaft, 2001, No. 3, pp. 126-136.

Kulatilaka, N.: The value of flexibility: A general model of real options, in: Trigeorgis, L. (editor): Real options in capital investment, Westport 1995, pp. 89-107.

Masten, S.E.: Instituional choice and the organisation of production: the make-or-buy decision, in: JITE, Vol. 142 (1986), No. 3, pp. 493-509.

McDonald, R./Siegel, D.: The value of waiting to invest, in: Quarterly Journal of Economics, Vol. 101 (1986), No. 4, pp. 707-727.

13

Meise, F.: Realoptionen als Investitionskalkül: Bewertung von Investitionen unter Unsicherheit, München 1998.

Micalizzi, A./Trigeorgis, L.: Project Evaluation, Strategy and Real Options, in: Trigeorgis, L.: Real Options and Business Strategy, London 1999, S. 1-19.

Pyndick, R. S.: Irreversibility, uncertainty and investment, in: Journal of Economic Literature, Vol. 23 (1991), No. 3, pp. 1110-1148.

Spence, Michael A.: Job Market Signaling, in: Quarterly Journal of Economics, Vol. 87 (1973), No. 3, pp. 355-374.

Stark, A. W.: Real options, (dis) investment, decision making and accounting measures of performance, in: Journal of Business Finance & Accounting, Vol. 27 (2000), No. 3/4, pp. 313-331.

Stiglitz, J.E.: Information and economic analysis, in: Parkin, M./Nobay, A.R. Current economic problems, Cambridge 1974, pp. 27-52.

Upton, D. M.: The management of manufacturing flexibility, in: California Management Review, Vol. 36 (1994), No. 2, p. 72-89.

Zerdick, A. et al.: Die Internet-Ökonomie, Berlin et al., 1999.

14

DAY 2

SESSION 2

Alexander Bukhvalov
"Application of Real Options to Strategic
Management in Transition Economies"

Application of Real Options to Strategic Management in Transition Economies

Alexander Bukhvalov

St.Petersburg State University

School of Management

Summary of Presentation to
WAENO International Real Option Workshop, Turku, 6-8 May, 2002

I cannot forecast to you the action of Russia.
It is a riddle wrapped in a mystery inside an enigma.
Winston Churchill

The notion is of real option is not a normative criteria for investment as NPV or IRR criterions are. Real options have been first invented by managers in their search for flexibility in risky environment, and only then they have been recognized and studied by academics. Whereas derivative instruments provide just passive hedging opportunity against risk exposure (in the sense that a hedger can do nothing if the market does not offer such an instrument), real options demand for managerial sophistication in projects design and, hence, they need active management style.

I would like methodologically to distinguish between economic analysis of investment under uncertainty and managerial decision-making in development of business projects in companies. As it is reasonably noticed by many authors (see, e.g., [Fernandez, 2001]) financial option valuations is often not appropriate for real option valuation. The reason is both in impossibility to replicate tracking portfolios and in difficulty to identify a suitable basic stochastic process with any well-defined probabilistic model (nothing to say about the geometric Brownian motion). I believe that wonderful techniques presented in [Dixit and Pindyck, 1993] is still adequate for economic and qualitative analysis of industries (see [Trigeorgis, 1996]. On the other hand, I do not think that it is appropriate techniques for real-life decision making in a company. There is now a bulk of important verbal papers on real options. I will mention just [Luerhman, 1998]. We know from the history of business that sophisticated methods have been never successful in their implementation and popular among managers. So the experts in the area should find a working compromise between sophistication of ideal models and obvious need to support decision making based on real options both qualitatively and quantitatively. I believe that simple binary tree approach is appropriate in justification of capital investment and strategic projects. I will use it in cases below.

In transition economies we face not only the usual valuation difficulties but we also see non-complete markets at the unusual level (few financial instruments, not more than 15-20 for Russia) and extraordinary uncertain market environment, where no fundamentals follow geometric Brownian or mean-reversion processes. Moreover, non-predictable business mentality, widespread of gray economy and opportunistic behavior make very difficult, if possible at all, any kind of forecast. Nevertheless, the real option approach gives easy explanations to many important economic events. Moreover, though real options are hardly known to majority Russian managers, they are widespread in actual development of strategies. I would like to thank Professor Frederick Balderston (UC, Berkeley) for many helpful discussions of "no forecast enterprise", a notion, which he has introduced in 2001 in his lectures to the students of the School of Management in St.Petersburg State University.

The paper consists of three sections.

Section 1 is devoted to explanation of recession in transition economies by analysis of major industries on the base of behavior of a typical business in the terms of real options. Section 2 studies Russian financial empires and asset stripping activities on the base of real options modeling. Section 3 contains some comparative information about the approach adopted by Russian and Finnish managers.

1. "Commerce vs Production" is the Dominant Real Option in Russia

Until now the derivatives market in Russia is slim even at the Russian scale (current FORTS trades incorporate 6 futures and 2 options contracts). This implies that passive hedging is available only for exporters of production who can use foreign derivative markets with the help from their foreign partners (oil and aluminum industries). Such Russian industries are price takers where the price is solved in the world market. This means that they face basically the same uncertainty as other international companies in the corresponding industries. Actually oil and natural gas, aluminum, and telecommunications (the latter enjoy the influence of globalization) are the most prosperous Russian industries. All of those can use and actually use real options build in these industries everywhere. Export industries in Russia are growing not only because of favorable oil prices but also because they are built-in usual international risk sharing.

During the 1992-97 the decline of Russian industrial output was equal to almost 50%. The reason was that domestically oriented manufactured industries faced with extreme uncertainty. Their production was not competitive at the world market and perspectives at the domestic market were not clear. The majority of these industries (not specially supported by the government as car-building industry) failed to operate in no-forecast environment. Since domestic demand should have been met by the market then Russian business has suspended domestic production and switched to the trade in imported goods.

Commerce always has an edge against production because of real option always built-in commercial activities. Commerce has higher flexibility because of possibility to change partners or to switch from foreign producers to domestic ones (as it took place in the food industry after August 1998 crisis) without high transformation costs. In industry (especially so technically weak and mentally rigid as Russian) it is very costly to use flexibility options as economies of scope. The switch from manufacturing to commerce, to a great deal, explains huge decline in GDP in Russia and other CIS countries in 1990s. On the contrary, the economic growth of China is explained by the fact that their manufacturing output has its niche in the world market (so they can orient to world prices, which are still volatile but more predictable than prices in transition economies).

Low level of FDI in Russia and CIS is analyzed in the same framework. A simple model of expected time to invest for a foreign investor is discussed.

Another issue is an unexpected gap between the high reputation of exact sciences in former Soviet Union and lack of technologically innovative ventures in Russia. In developed economies if innovation is of no forecast type then venture and angel capital are used. Though a single unit still preserves no forecast feature the situation is under control. Venture firm actually sells a call option on their project (if business fails she has no debt obligations; investor's yield is unknown and uncertain since it is just a share of profit of a successful venture business). Venture capitalist has a pool of these call options bought. It diversifies her risk. Flexibility is expensive (cf. Netscape case). So, usually venture business does not obey much flexibility. Nevertheless, flexibility exists for market as a whole. Projects devoted to production of similar products, which differ in some details and technology will be supported different venture capitalists. Hence, instead of one firm with build-in flexibility we have a number of rigid venture projects financed from different sources. There are no venture capitalists in Russia. Probably it is partially a path dependency feature since previously engineers used to perform in stable environment of state enterprises. In any case this real options portfolio is not in use in Russia.

2. Asset Stripping as Costs Minimizing Behavior: Performance of Business Groups in Russia

Another interesting issue is an explanation of behavior of so-called Russian financial empires (financial-industrial groups) in 1990s. A simple binary tree model of call option valuation shows that conglomerate groups have many advantages against standalone enterprises because of asset stripping opportunities. Interros group and Menatep Bank group give us instructive cases of such optimal behavior. Both groups used pocket bank technology during 1998 crisis. This can be also treated in the framework of new institutional economics (transaction costs and analysis of contracts types). Certainly, we come here to the issue of poor corporate governance.

Opportunistic behavior of top managers is one of major threats both for national economy and shareholders. Asset stripping, off-shore operation, price manipulation and many other techniques are used. This undermines the value of Russian companies in the eyes of both Russian and foreign investors. Sindanco/BP Amoco case is analyzed as a real option to wait for a transnational corporation investing in Russia.

Let me to mention that opportunistic behavior options are widespread in Western economies as well. For example, we can treat the behavior of Enron Corp's top managers as a complex real option with the disclosure of fraud by the society as basic variable.

3. What is Different in Treatment of Real Options in Russia and Finland?

This part is based on the analysis of essay works on real options, which were given by the author to the students of full-time MBA program and executive MBA program in St.Petersburg State University (2001/2002), on one hand, and Ph.D. students in finance in Finland (2000/2001, nation-wide program; course was delivered in HANKEN). The last topic looks very different from two previous ones but it gives some support to the idea that in both countries managers are enthusiastic about the use of options but the scale of projects proposed by Finnish students is usually much greater and more oriented to productive activities rather than commerce. Managers at MBA program (mostly top managers from small and medium size enterprises in St.Petersburg area) also often mention opportunistic real options in the behavior of owners and enterprises related to the gaps in legislation and enforcement.

References

Dixit, A.K. and R.S. Pindyck (1994), *Investment under Uncertainty*, Princeton Univ. Press.

Fernadez, P. (2001), *Valuing real options: frequently made errors*, SSRN.

Luehrman, T.A. (1998), *Strategy as Portfolio of Real Options*, Harvard Business Review, September-October, pp. 89-99.

Trigerogis, Lenos (1996), *Real Options*, Cambridge, MA: MIT Press.

Alexander Bukhvalov
Dept. Finance
School of Manangement
St.Petersburg State University
per. Dekabristov, 16
St.Petersburg 199155 Russia
Fax: +7 (812) 3500406
E-mail: bukh@pop3.rcom.ru

DAY 2

SESSION 3

Christer Carlsson and Robert Fullér
"Fuzzy Real Option Valuation –
A Breakthrough Theory"

Mikael Collan
"Investment Decisions and Fuzzy Numbers"

Péter Majlender
"Optimal Timing for the Exercise of Real Options"

Markku Heikkilä
"User Interfaces and Knowledge Representation in
Real Options"

Investment Decisions and Fuzzy Numbers

Mikael Collan

IAMSR

07.05.2002   Mikael Collan
                                    IAMSR

The Setting

- Very large investments have a long economic life (10-60 years)

- This means that it is increasingly difficult to estimate future cash flows - methods using precise estimates are misleading

- Using fuzzy numbers is a comprehensive way of representing the uncertainty

07.05.2002   Mikael Collan
                                    IAMSR

The Setting

- For example, an investment in a nuclear power plant has an economic life of about 60 years

- There is considerable uncertainty about electricity prices, which makes it highly unrealistic to present precise estimates of the future cash flows

07.05.2002   Mikael Collan
                                    IAMSR

State of the Art
Selected References

- Buckley (1987) - Fuzzy FV and PV

- Li Calzi (1990) - Discussion about a general setting

- Buckley (1992) - Fuzzy IRR

- Kuchta (2000) - 7 different fuzzy methods

- Tarrazo and Gutierrez (2000) - Discussion about economic expectations and fuzzy sets in finance

07.05.2002   Mikael Collan
                                    IAMSR

State of the Art
Selected References

- Muzzioli and Torricelli (2000) - Fuzzy binomial option pricing

- Carlsson and Fullér (2000a) - Fuzzy real option valuation

- Zmeskal (2001) - Firm valuation as an option with fuzzy-stochastic methodology

07.05.2002　　　　　Mikael Collan
IAMSR

Important Points

- The fuzzy methods are original constructions - they are not "fuzzifications" of existing constructions

- When cash flow expectations are given as fuzzy numbers information is not lost by compromises in finding a single correct number

07.05.2002　　　　　Mikael Collan
IAMSR

Fuzzy Cash Flow Estimates

- Fuzzy estimates of future cash flows give a correct presentation of the uncertainty involved
- If there are more than one expert opinion they can be included in a single fuzzy number (must state lower & upper possible values [a,b]), Bardossy et al. (1993)
- There are methods that enable inclusion of trend information into fuzzy numbers

07.05.2002   Mikael Collan
                                    IAMSR

Results from Fuzzy Methods

- Results of fuzzy investment planning are also fuzzy numbers
- The question is: " What does the fuzzy result tell us?"
- A fuzzy result includes within it sensitivity with respect to uncertainty, as understood by the expert

07.05.2002   Mikael Collan
                                    IAMSR

Problems with Interpretation

- If the fuzzy result is very wide it may be hard to get a good insight on the meaning of the result

- For better interpretations the result can be reduced with a *vertical and horizontal reduction of uncertainty*

Result =

07.05.2002   Mikael Collan
IAMSR

Vertical Reduction of Uncertainty

- Vertical reduction: The fuzzy number is reduced so that points with a membership value of less than α ($\alpha \in [0,1]$) will be neglected; this is called α-cut

- The possibilistic variance of the fuzzy number is reduced, Carlsson and Fullér (2001)

07.05.2002   Mikael Collan
IAMSR

Vertical Reduction of Uncertainty

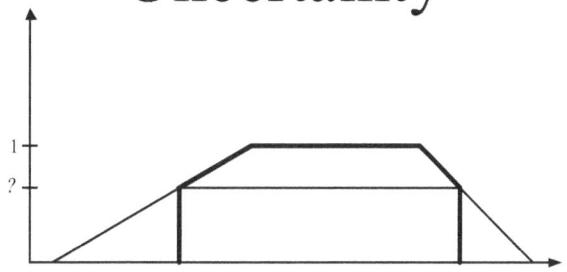

Effect of the vertical reduction of uncertainty

This method can be used when the "sides" of the fuzzy number are large (or very large) and they need to be reduced.

07.05.2002   Mikael Collan
                               IAMSR

Horizontal Reduction of Uncertainty

- Horizontal reduction of uncertainty can be used when the core of the fuzzy number is very wide

- The method reduces the possibilistic variance (in proportion to α) of the fuzzy number, but the mean value remains the same.

07.05.2002   Mikael Collan
                               IAMSR

Horizontal Reduction of Uncertainty

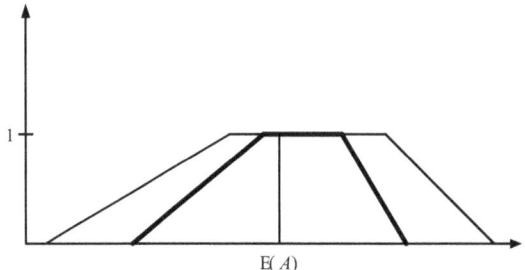

Effect of the horizontal reduction of uncertainty

07.05.2002
Mikael Collan
IAMSR

Reduction of Uncertainty

- It is possible to use both methods, the vertical and the horizontal reduction simultaneously

- The same level of α can be applied to both methods

- The resulting fuzzy number has a narrower core and shorter sides

07.05.2002
Mikael Collan
IAMSR

Further Problems with Interpretation

- If we have multiple projects, from among which the best must be selected we need to rank fuzzy numbers

- This is not as straightforward as with crisp numbers, because fuzzy numbers can be overlapping.

- Some special methods are needed

07.05.2002   Mikael Collan
IAMSR

Ranking Fuzzy Numbers

- The fuzzy numbers can be defuzzified to a crisp number and then ranked

- A value function can be used to give a value to the fuzzy numbers and then they can be ranked according to the values (Carlsson et al. 2001)

- There are a number of different methods

07.05.2002   Mikael Collan
IAMSR

Suggestions

- Using the fuzzy cash flow estimates in yearly budgeting in investment projects
- Using results from fuzzy capital budgeting methods as a risk management tool -
- Using real option value and fuzzy real option value to find the maximum negative potential of an investment

07.05.2002   Mikael Collan
IAMSR

Positive + Negative potential = Total Flexibility

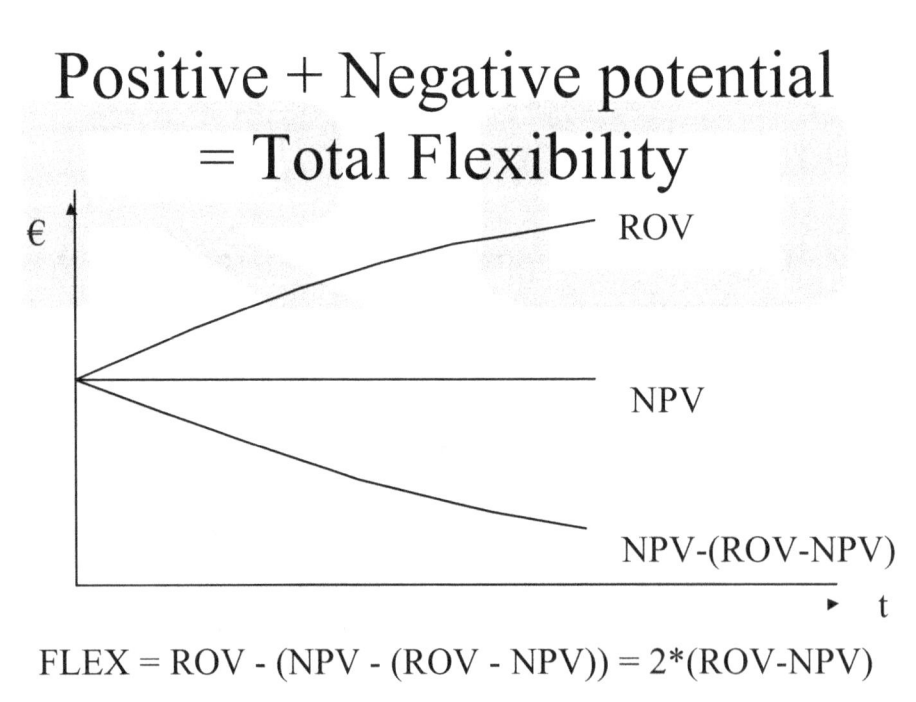

$$FLEX = ROV - (NPV - (ROV - NPV)) = 2*(ROV-NPV)$$

07.05.2002   Mikael Collan
IAMSR

References

Bardossy, A., Duckstein, L., and Bogardi, I., 1993, Combination of fuzzy numbers representing expert opinions, Fuzzy Sets and Systems, 57, pp. 173-181

Carlsson, C. and Fullér, R., 1999, Capital budgeting problems with fuzzy cash flows, Mathware and Soft Computing, 6, pp. 81-89

Carlsson, C. and Fullér, R., 2000a, On fuzzy real option valuation, TUCS - Turku Centre for Computer Science, Technical Report No. 367 - available at www.tucs.fi

Carlsson, C. and Fullér, R., 2000b, Real option evaluation in fuzzy environment, Proceedings of the International Symposium of Hungarian Researchers on Computational intelligence, Budapest Polytechnic, pp. 69-77

Carlsson, C. and Fullér, R., 2001, On possibilistic mean value and variance of fuzzy numbers, Fuzzy Sets and Systems, 122, pp. 315-326

Carlsson, C., Fuller, R., and Majlender, P., 2001, Project selection with fuzzy real options, Proceedings of the Second International Symposium of Hungarian Researchers on Computational Intelligence, pp. 81-89

Collan, M., Carlsson, C., and Majlender, P., forthcoming, Fuzzy Black and Scholes real options pricing, Presented at the 12th MiniEURO conference in Brussels, April 2002

Collan, M. and Majlender, P., forthcoming, A method for including dynamic trend information in fuzzy pricing of real options, Presented at 12th MiniEURO conference in Brussels, April 2002

07.05.2002   Mikael Collan
                              IAMSR

References

Kuchta, D., 2000, Fuzzy capital budgeting, Fuzzy Sets and Systems, 111, pp. 367-385

Li Calzi, M., 1990, Towards a general setting for the fuzzy mathematics of finance, Fuzzy Sets and Systems, 35, pp. 265-280

Muzzioli, S. and Torricelli, C., 2000, Combining the theory of evidence with fuzzy sets for binomial option pricing, materiale di discussione n. 312, Dipartimento di Economia Politica, Università degli Studi di Modena e Reggio Emilia, May

Tarrazo, M. and Gutierrez, L., 2000, Exonomic expectations, fuzzy sets and financial planning, European Journal of Operational Research, 126, pp. 89-105

Zmeskal, Z., 2001, Application of the fuzzy-stochastic methodology to appraising the firm value as a European call option, European Journal of Operational Research, 135, pp. 303-310

07.05.2002   Mikael Collan
                              IAMSR

Appendix 1:

Vertical reduction of uncertainty - possibilistic variance

Using the definition of a possibilistic variance (Carlsson and Fullér 2000b)

$$\mathrm{Var}(A) = \frac{1}{2}\int_0^1 [a_2(\gamma) - a_1(\gamma)]^2 \gamma d\gamma$$

We find that the possibilistic variance of the original fuzzy number is larger:

$$\mathrm{Var}(B) = \frac{1}{2}\int_0^1 [b_2(\gamma) - b_1(\gamma)]^2 \gamma d\gamma = \frac{1}{2}\int_0^\alpha [b_2(\gamma) - b_1(\gamma)]^2 \gamma d\gamma + \frac{1}{2}\int_\alpha^1 [b_2(\gamma) - b_1(\gamma)]^2 \gamma d\gamma$$

$$= \frac{1}{2}\int_0^\alpha [a_2(\alpha) - a_1(\alpha)]^2 \gamma d\gamma + \frac{1}{2}\int_\alpha^1 [a_2(\gamma) - a_1(\gamma)]^2 \gamma d\gamma$$

$$\leq \frac{1}{2}\int_0^\alpha [a_2(\gamma) - a_1(\gamma)]^2 \gamma d\gamma + \frac{1}{2}\int_\alpha^1 [a_2(\gamma) - a_1(\gamma)]^2 \gamma d\gamma$$

$$= \frac{1}{2}\int_0^1 [a_2(\gamma) - a_1(\gamma)]^2 \gamma d\gamma = \mathrm{Var}(A).$$

07.05.2002   Mikael Collan
IAMSR

Appendix 2:

Horizontal reduction of uncertainty

Using the definition of the mean value [Carlsson and Fullér (2001b)]

$$\mathrm{E}(A) = \int_0^1 [a_1(\gamma) + a_2(\gamma)]\gamma d\gamma$$

we have that

$$\mathrm{E}(B) = \int_0^1 [b_1(\gamma) + b_2(\gamma)]\gamma d\gamma = \int_0^1 [\alpha(a_1(\gamma) + a_2(\gamma)) + 2(1-\alpha)\mathrm{E}(A)]\gamma d\gamma$$

$$= \alpha\int_0^1 [a_1(\gamma) + a_2(\gamma)]\gamma d\gamma + (1-\alpha)\mathrm{E}(A) = \alpha\mathrm{E}(A) + (1-\alpha)\mathrm{E}(A) = \mathrm{E}(A),$$

07.05.2002   Mikael Collan
IAMSR

Appendix 3:
Horizontal reduction of uncertainty

using the definition of a possibilistic variance

$$\text{Var}(A) = \frac{1}{2}\int_0^1 [a_2(\gamma) - a_1(\gamma)]^2 \gamma d\gamma$$

we find that

$$\text{Var}(B) = \frac{1}{2}\int_0^1 [b_2(\gamma) - b_1(\gamma)]^2 \gamma d\gamma = \frac{1}{2}\int_0^1 [\alpha a_2(\gamma) - \alpha a_1(\gamma)]^2 \gamma d\gamma$$

$$= \alpha^2 \frac{1}{2}\int_0^1 [a_2(\gamma) - a_1(\gamma)]^2 \gamma d\gamma = \alpha^2 \text{Var}(A).$$

07.05.2002

Mikael Collan
IAMSR

Dynamic Decision Tree – Optimal Timing for the Exercise of Real Options

Péter Majlender

Abstract

We shall represent strategic planning problems by dynamic decision trees, in which the nodes are projects that can be deferred or postponed for a certain period of time.

The Problems

Having an investment opportunity how much is it worth right now?

or

Having a deferrable project how long should it be postponed to gain maximal profit?

or

Having deferrable projects which are related to each other how should they be phased and scheduled to gain maximal profit in the aggregate?

Set-up

- Step 1: Investment opportunity of type now or never

- Step 2: Investment opportunity with deferral flexibility

- Step 3: Investment opportunities which are related to each other

Generalization →

Tools

- Step 1 – Net Present Value Analysis
  Using NPV table

- Step 2 – Black-Scholes Theory
  Using Black-Scholes pricing formula

- Step 3 – Dynamic Decision Tree
  Using theory of compound options

Step 1 – NPV Analysis

Consider the NPV table of the project

r Risk-adjusted Discount Rate
10.00%

t Time (in years)	V Expected Revenues (in millions)	x Investment Cost (in millions)
0	€ 100.00	€ 750.00
1	€ 115.00	
2	€ 127.00	
3	€ 138.00	
4	€ 148.00	
5	€ 156.00	

$$NPV = S_0 - X$$

Data needed:

S_0 – PV of Expected Revenues

X – Investment Cost

r – Required rate of return on the project

$$S_0 = \sum_{t=0}^{L} \frac{V_t}{(1+r)^t}$$

where V_t denotes the expected cash flows at time t

Step 1 – Fuzzy Approach

Usually, the expected cash flows in the NPV table cannot be characterized by numbers

These quantities can be estimated by using possibility distributions, i.e. fuzzy numbers

Consider

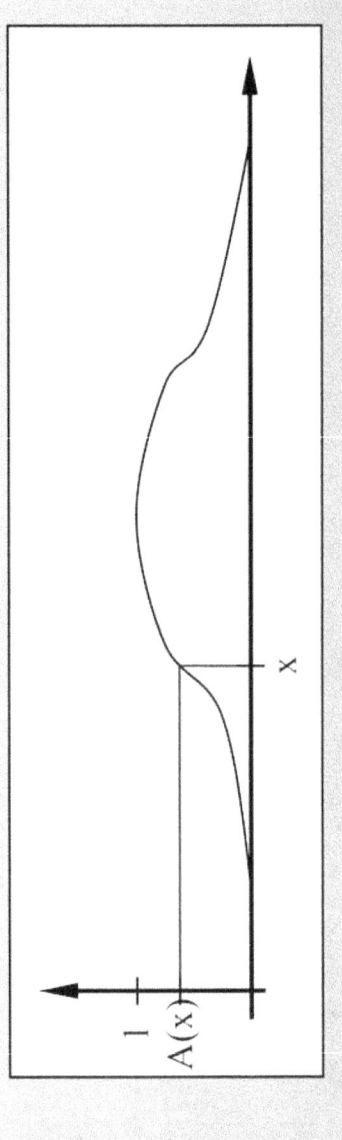

$A(x)$: the degree of possibility of the statement "x is the value of the expected cash flow"

Step 1 – Fuzzy Approach

The normal numbers can be substituted by possibilistic distributions in the NPV table

It is sufficient to consider _trapezoidal fuzzy numbers of the form_

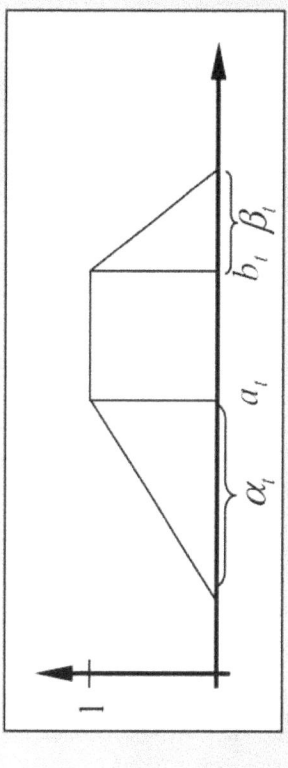

$$V_t = (a_t, b_t, \alpha_t, \beta_t) \qquad t = 0,1,\ldots,L$$

the value of the expected revenues at year t of the project is approximately in the interval $[a_t, b_t]$

$$X = (x_1, x_2, \alpha', \beta')$$

- They can be characterized and referred by only 4 parameters
- The calculations with them are straightforward
- They have the ability to be able to characterize the considered quantities – Managers like using them

Step 1 – Fuzzy Approach

Now the following Fuzzy NPV table is obtained

r
Risk-adjusted Discount Rate

10.00%

t Time (in years)	V Expected Revenues (in millions)				X Investment Cost (in millions)			
	a	b	α	β	a	b	α	β
0	€95.00	€105.00	€5.00	€5.00	€740.00	€760.00	€15.00	€15.00
1	€108.00	€120.00	€8.00	€5.00				
2	€125.00	€132.00	€12.00	€7.00				
3	€135.00	€142.00	€15.00	€5.00				
4	€145.00	€151.00	€4.00	€4.00				
5	€150.00	€160.00	€10.00	€11.00				

$$V_0, V_1, ..., V_L \longrightarrow \quad S_0 = \sum_{t=0}^{L} \frac{V_t}{(1+r)^t}$$

$$FNPV = S_0 - X$$

Step 2 – Black-Scholes Theory

The price of a financial European option is typically estimated by the application of the Black-Scholes formula:

$$C_T = S_0 e^{-\delta T} N(d_1) - X e^{-r_f T} N(d_2),$$

$$d_1 = \frac{\ln(S_0 / X) + (r_f - \delta + \sigma^2 / 2)T}{\sigma \sqrt{T}}$$

$$d_2 = d_1 - \sigma \sqrt{T}$$

The value of an investment opportunity, which has to be postponed T years can be calculated by using this formula

Step 2 – Black-Scholes Theory

Notations:

$NPV(r) \begin{cases} S_0 \\ X \end{cases}$

S_0	–	Present value of expected cashflows
X	–	Present value of fixed costs
σ	–	Uncertainty of expected cashflows
T	–	Time to expiry
δ	–	Value lost over duration of option
r_f	–	Risk-free interest rate
$N(\cdot)$	–	Cumulative normal distribution function

Step 2 – Fuzzy Approach

Considering the Fuzzy NPV table the data gained from it can be input to the Black-Scholes formula

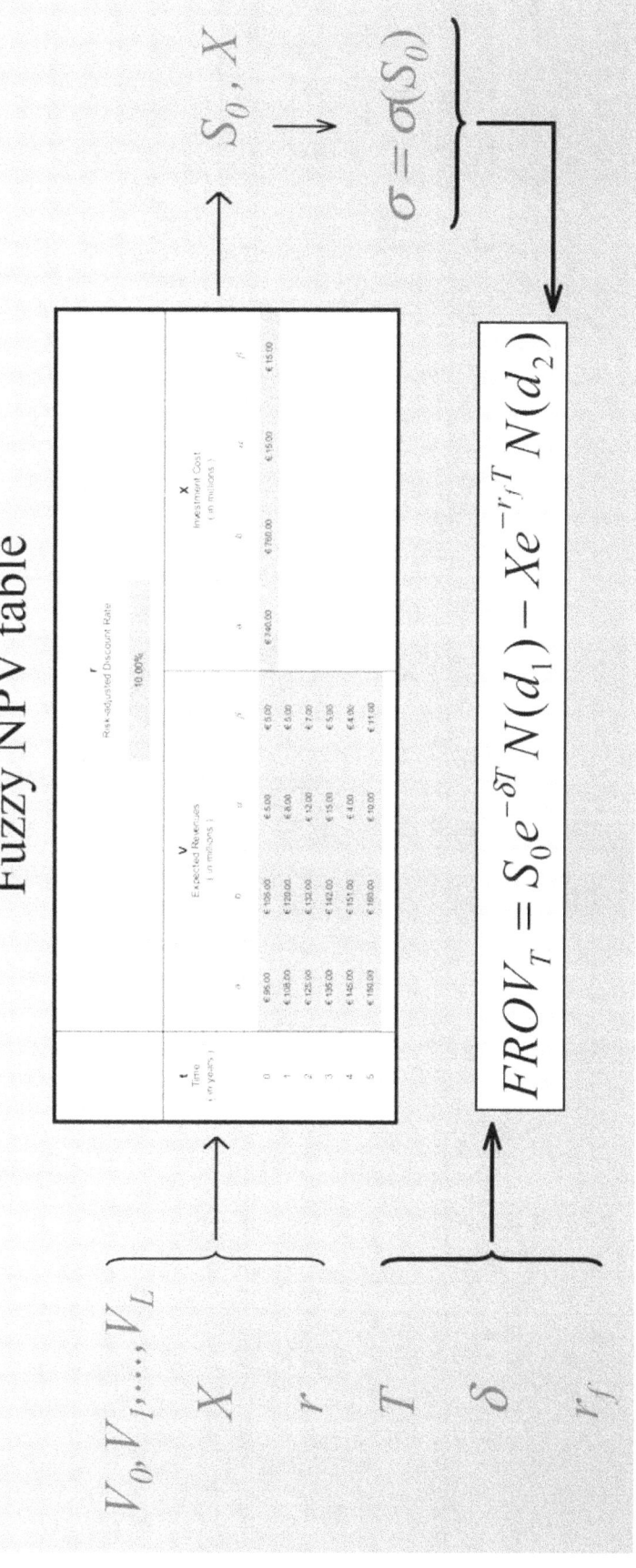

Fuzzy NPV table

$$FROV_T = S_0 e^{-\delta T} N(d_1) - X e^{-r_f T} N(d_2)$$

Step 2 – Fuzzy Approach

The fuzzy numbers in the Fuzzy NPV table implicitly contain the uncertainty of future cashflows

$$\sigma(S_0) = \sqrt{\left(\frac{s_2 - s_1}{2} + \frac{\alpha + \beta}{6}\right)^2 + \frac{(\alpha + \beta)^2}{72}}$$

$$S_0 = (s_1, s_2, \alpha, \beta)$$

$$\sigma = \sigma(S_0)$$

The possibilistic variance of the present value of expected cashflows

The most possible cashflows in the future can be characterized by the possibilistic mean value of the present value of expected cashflows

$$E(S_0) = \frac{a + b}{2} + \frac{\beta - \alpha}{6}$$

$$E(S_0)$$

Step 2 – Fuzzy Approach

Characterizing the most probable investment

costs as

$E(X)$

$$E(X) = \frac{x_1 + x_2}{2} + \frac{\beta' - \alpha'}{6}$$

$$X = (x_1, x_2, \alpha', \beta')$$

we have the following formula for computing

fuzzy real option values

$$FROV_T = S_0 e^{-\delta T} N(d_1) - X e^{-rT} N(d_2),$$

$$d_1 = \frac{\ln(E(S_0)/E(X)) + (r_f - \delta + \sigma^2/2)T}{\sigma\sqrt{T}}$$

$$d_2 = d_1 - \sigma\sqrt{T}$$

Step 2 – Fuzzy Approach

Having an investment opportunity, which <u>can be deferred up to T years</u>, its value can be obtained as

$$FROV_{[0,T]} = \max_{0 \le t \le T} FROV_t$$

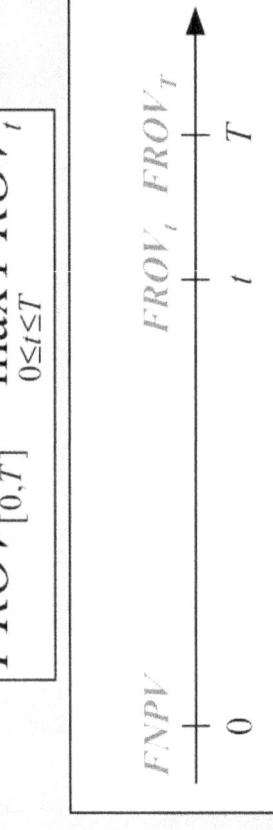

We should enter into the project in time t^*, where

$$FROV_{t^*} = \max_{0 \le t \le T} FROV_t$$

Obviously, for different deferral time t there can be different cashflows and investment costs estimates, i.e. different Fuzzy Fuzzy NPV tables

Step 2 – Fuzzy Approach

In the following we assume that any project can be deferred only whole years

(Time discretisation)

$$FROV_{[0,T]} = \max_{t=0,1,\dots,T} FROV_t$$

There exists a developed Excel platform, in which all of this theory are implemented

Microsoft Excel
Workbook

In the program there are some samples how each Fuzzy NPV table belonging to some allotted deferral time can be obtained

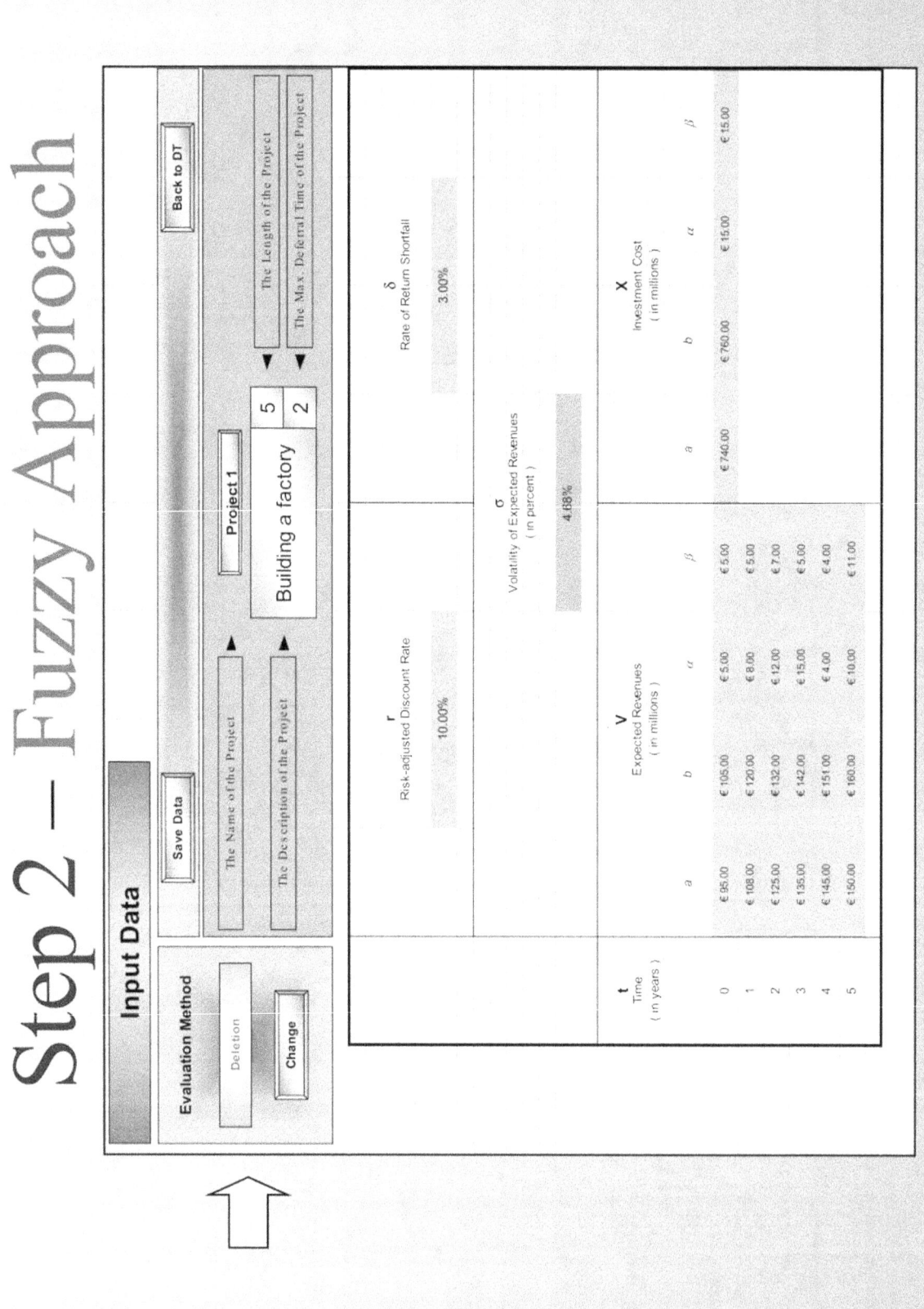

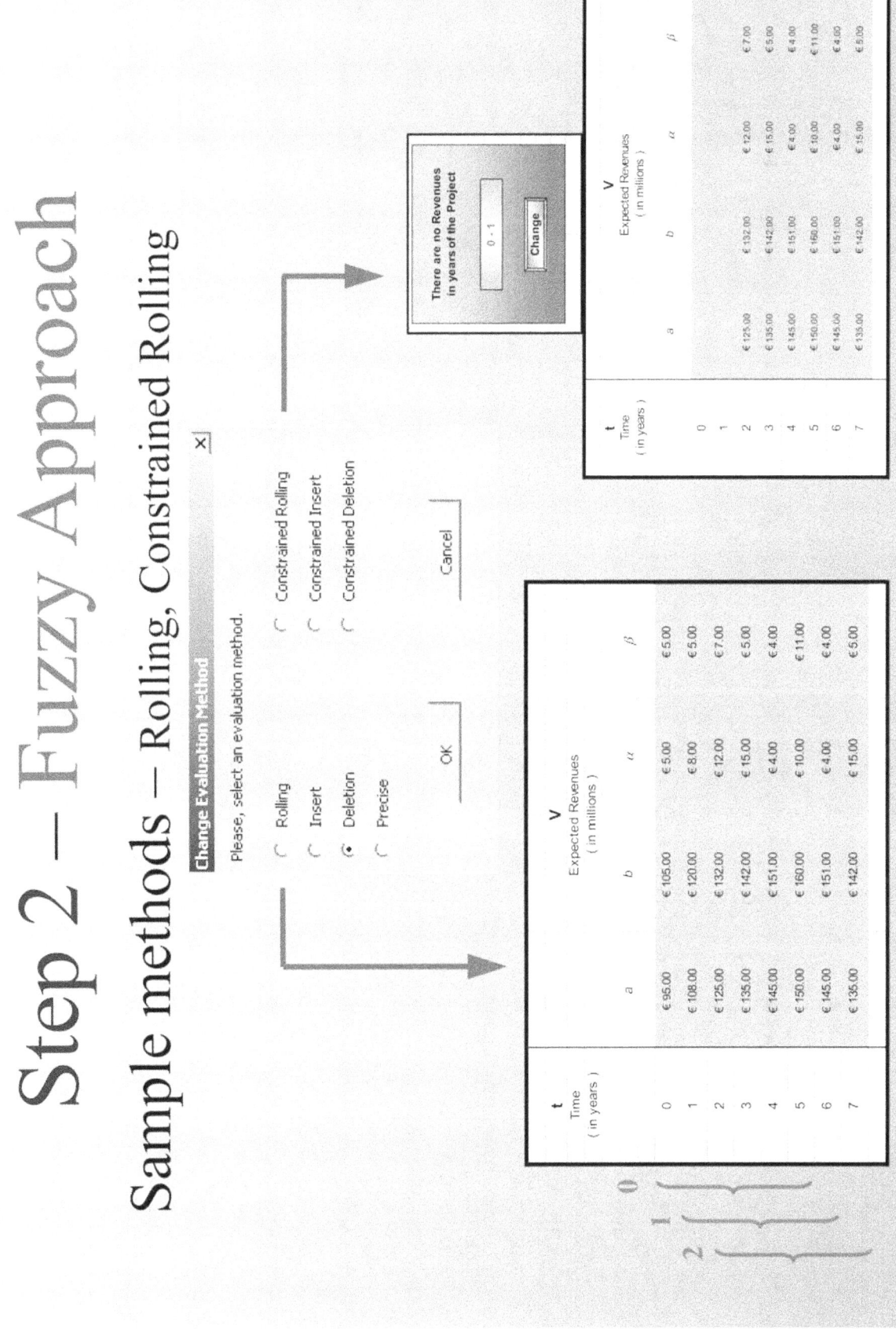

Step 2 – Fuzzy Approach

Sample methods – Rolling, Constrained Rolling

Step 2 – Fuzzy Approach

Sample methods – Insert, Constrained Insert

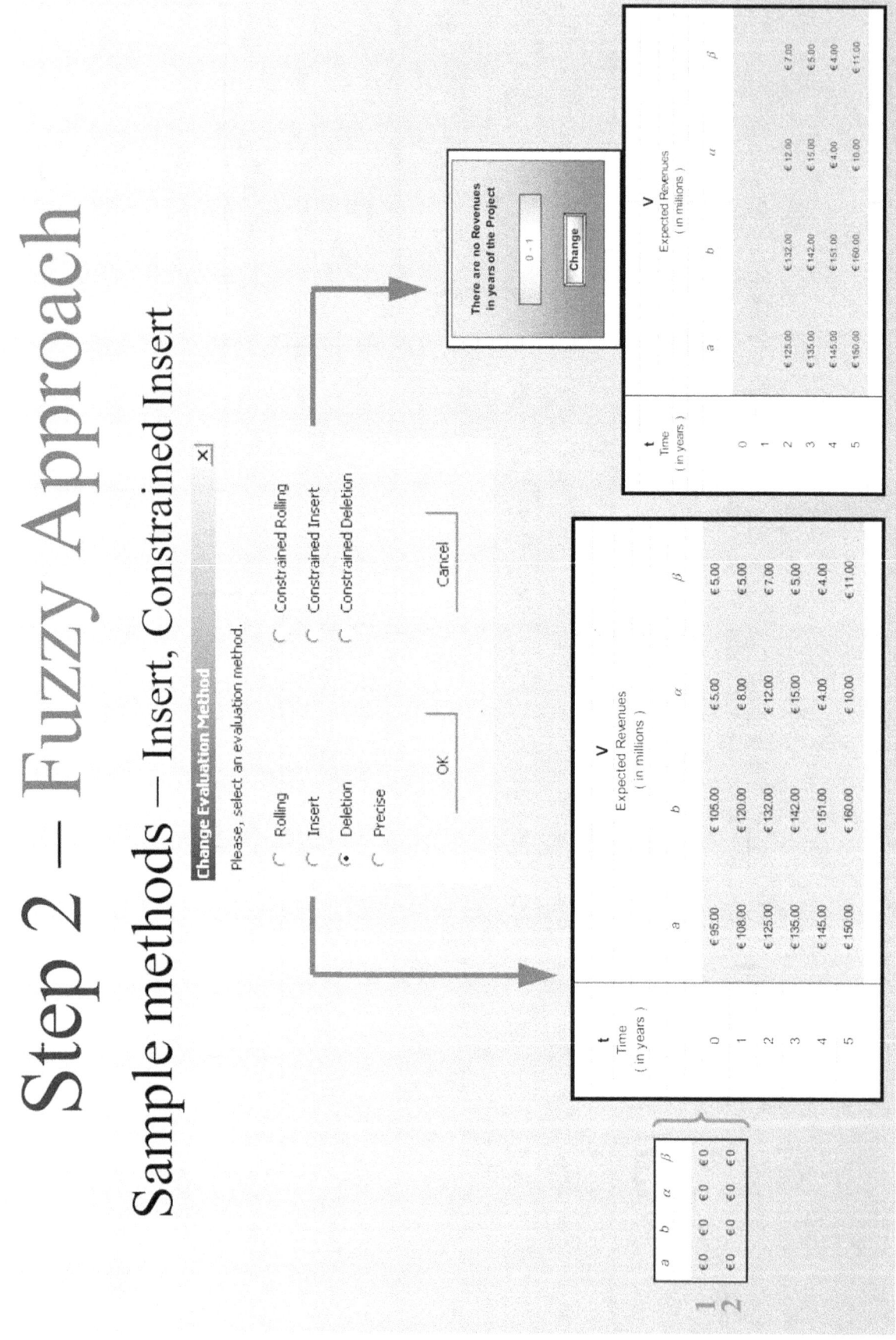

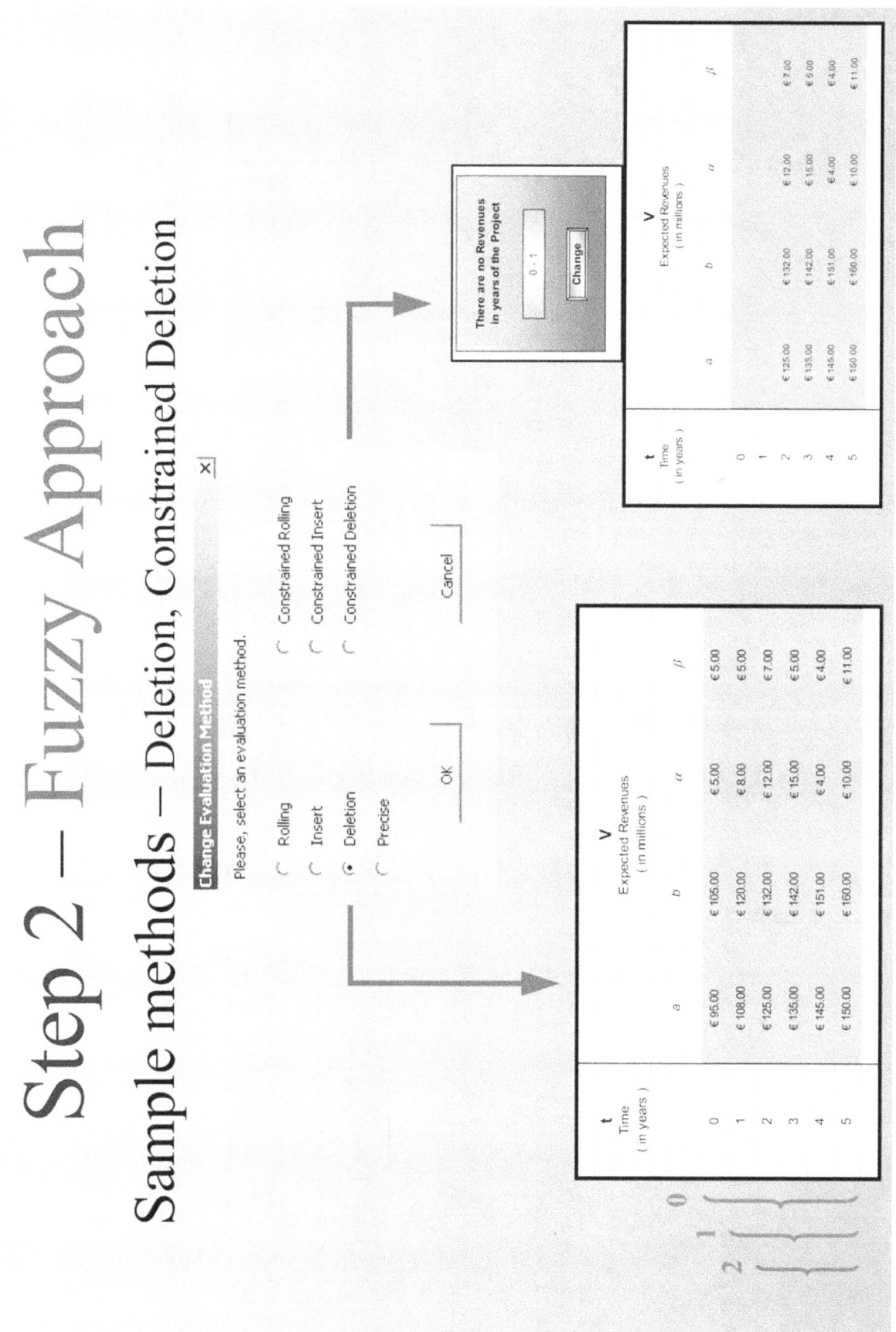

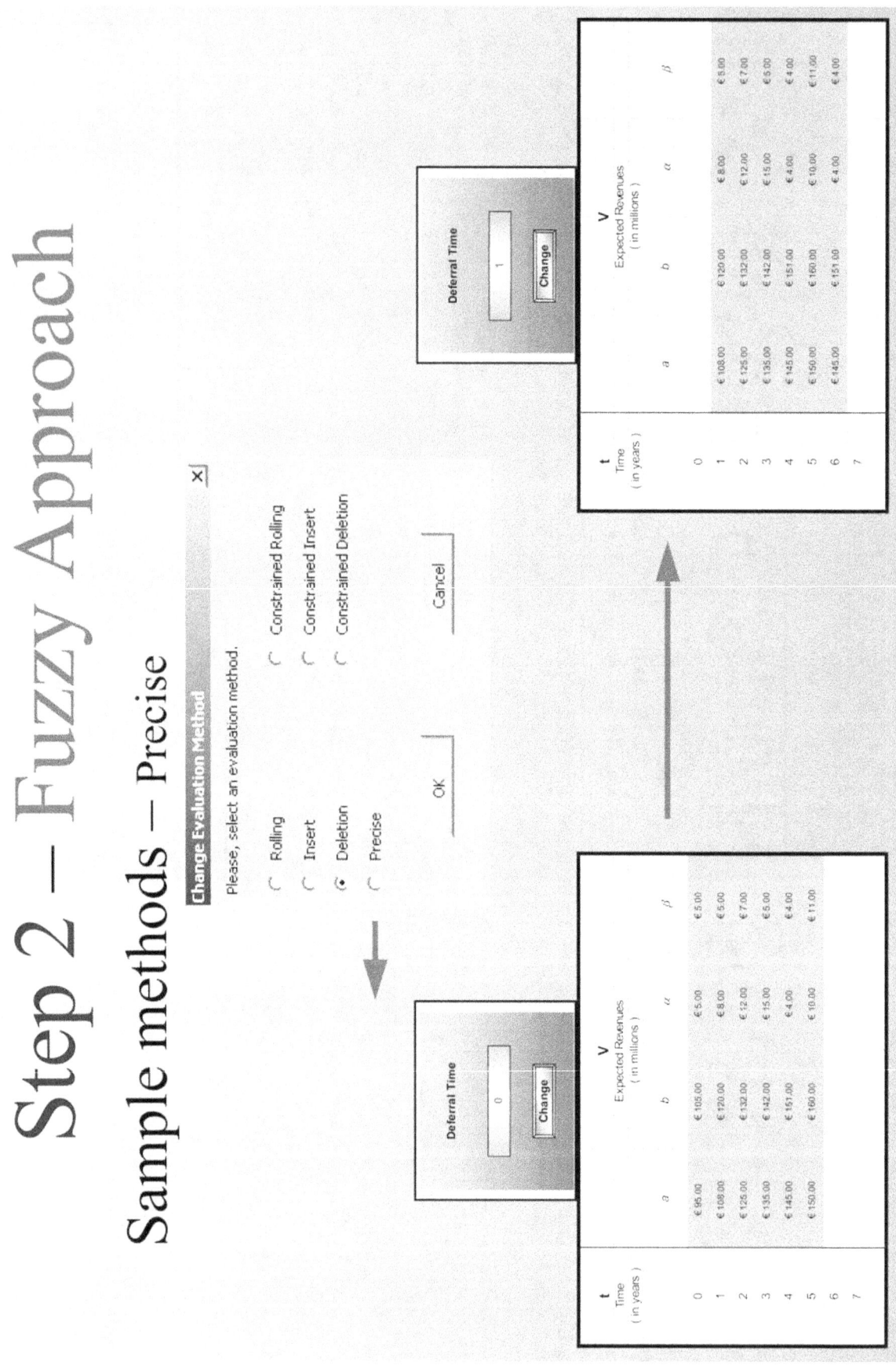

Step 3 – Dynamic Decision Tree

Decision trees are excellent tools for making financial decisions where a lot of vague information needs to be taken into account

Phasing and scheduling of projects which are related to each other can make a huge impact on the value of that set of projects

Decision trees provide an effective structure in which alternative decisions and the implications of taking those decisions can be laid down and evaluated

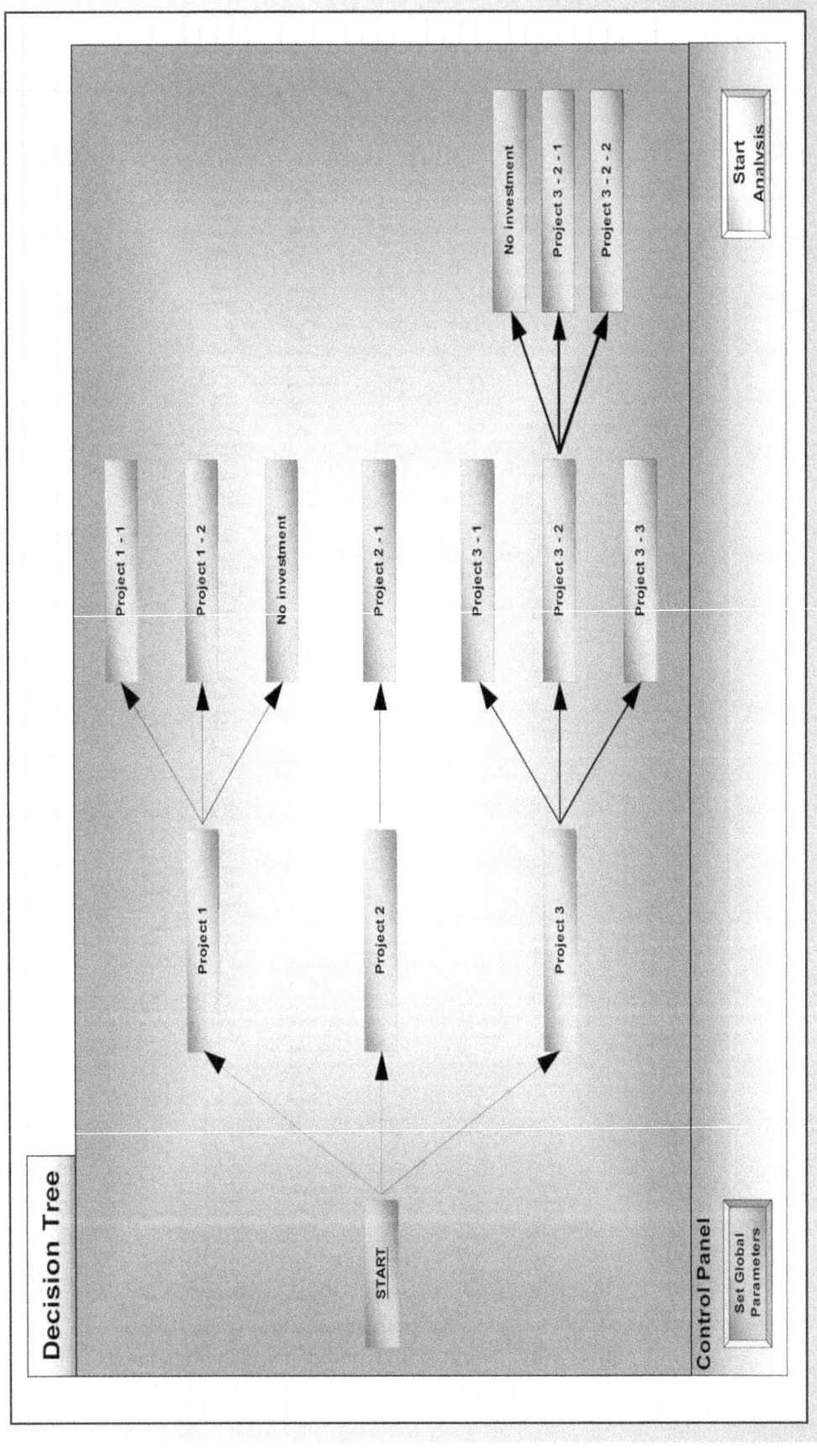

Step 3 – Dynamic Decision Tree

Step 3 – Dynamic Decision Tree

By phasing and scheduling projects, every step in a project opens or closes the possibility for further options

This is called a chain of growth options, or a compound growth option

The nodes of the tree are projects, which can be deferred or postponed for a certain period of time

Project 1

Step 3 – Dynamic Decision Tree

Project 1

Input Data

Project 1

Length of the Project :	5
Maximum Deferral Time of the Project :	3

Control Panel

Save Data | Back to DT

r
Risk-adjusted Discount Rate
10.00%

δ
Rate of Return Shortfall
0.00%

σ
Volatility of Expected Revenues
5.29%

X
Investment Cost
(in millions)

a	b	α	β
€ 740.00	€ 760.00	€ 15.00	€ 15.00

V
Expected Revenues
(in millions)

t Time (in years)	a	b	α	β
0	€ 95.00	€ 105.00	€ 5.00	€ 5.30
1	€ 108.00	€ 120.00	€ 8.00	€ 5.30
2	€ 125.00	€ 132.00	€ 12.00	€ 7.00
3	€ 135.00	€ 142.00	€ 15.00	€ 5.00
4	€ 145.00	€ 151.00	€ 4.00	€ 4.00
5	€ 150.00	€ 160.00	€ 10.00	€ 11.00
6	€ 145.00	€ 151.00	€ 4.00	€ 4.00
7	€ 135.00	€ 142.00	€ 15.00	€ 5.00
8	€ 130.00	€ 138.00	€ 10.00	€ 8.00

Step 3 – Dynamic Decision Tree

Creating options can buy us time to think and gain information to decide whether or not go ahead with a certain bigger investment

The decision rule has to be reapplied every time new information arrives during the deferral period to see how the optimal investment strategy might change in light of the new information

Step 3 – Dynamic Decision Tree

Using the theory of real options we have been able to identify the optimal path of the tree, i.e. the path with the biggest real option value in the end of the planning period

Step 3 – Dynamic Decision Tree

Example:

Year 0: Decision: Enter Project 3 after 2 years

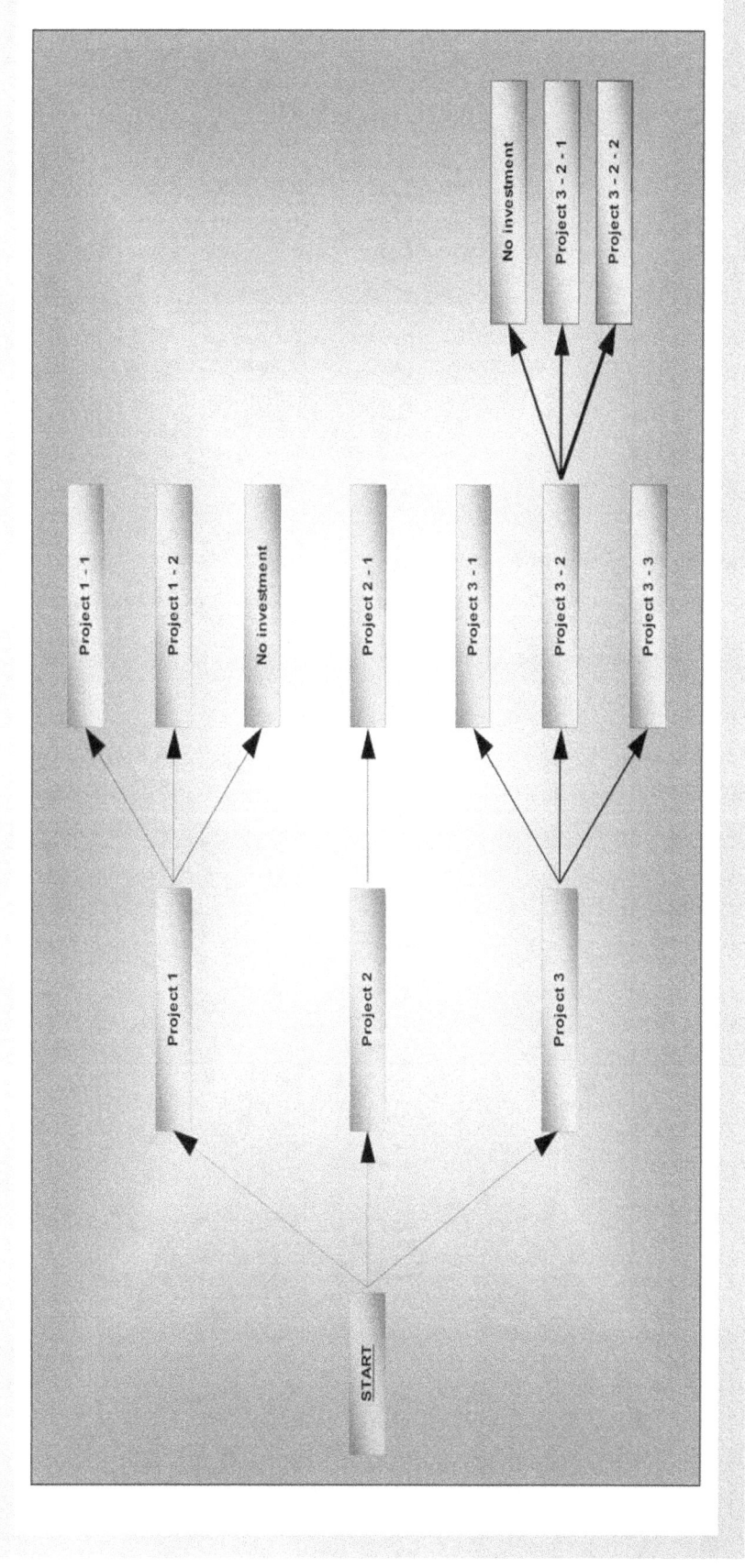

Step 3 – Dynamic Decision Tree

Year 2: Decision: Enter Project 3-2 after 1 year

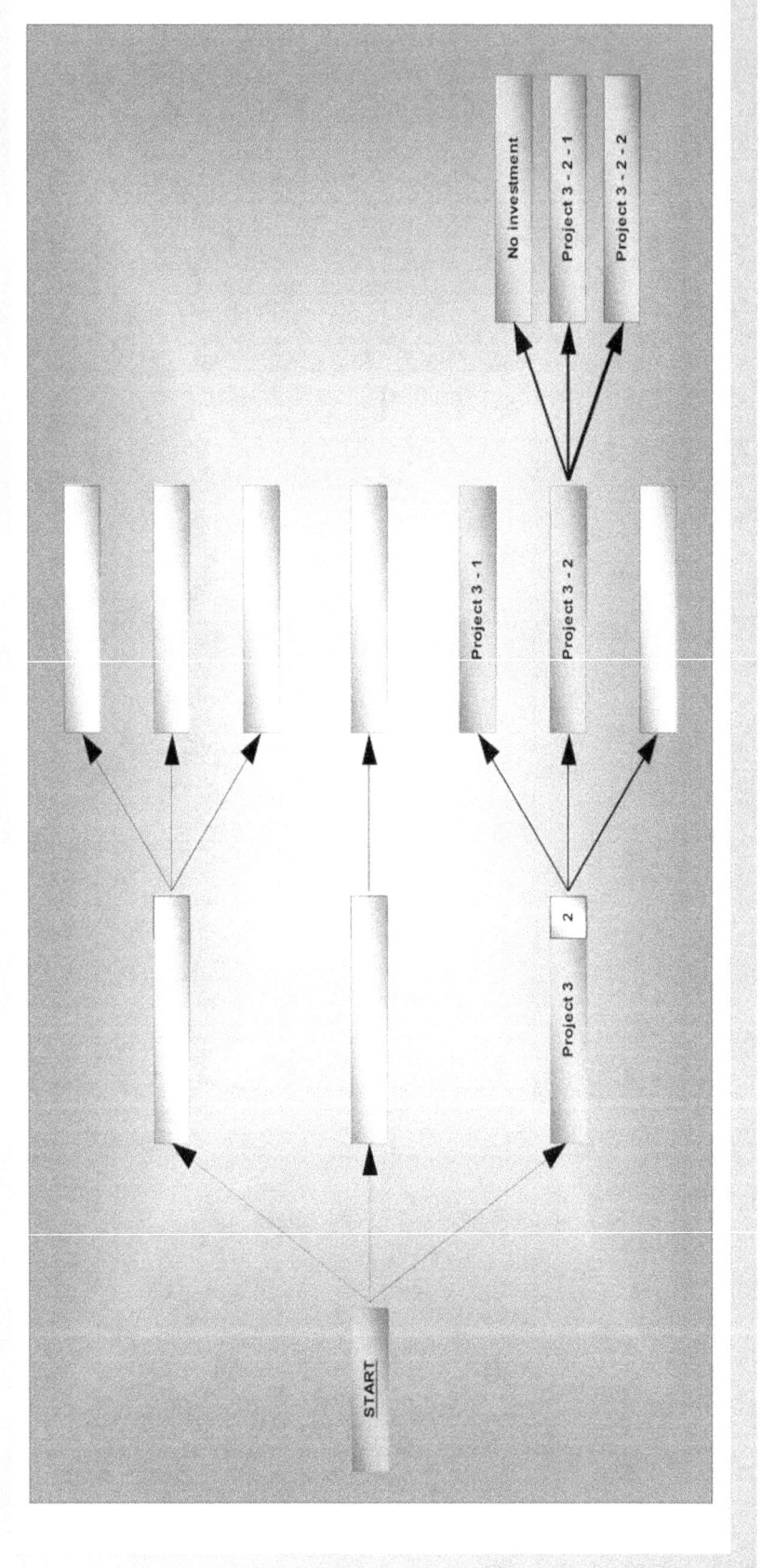

Step 3 – Dynamic Decision Tree

Year 3: Decision: Enter Project 3-2-2 after 3 years

Step 3 – Dynamic Decision Tree

Year 6: Result: Optimal path of the tree

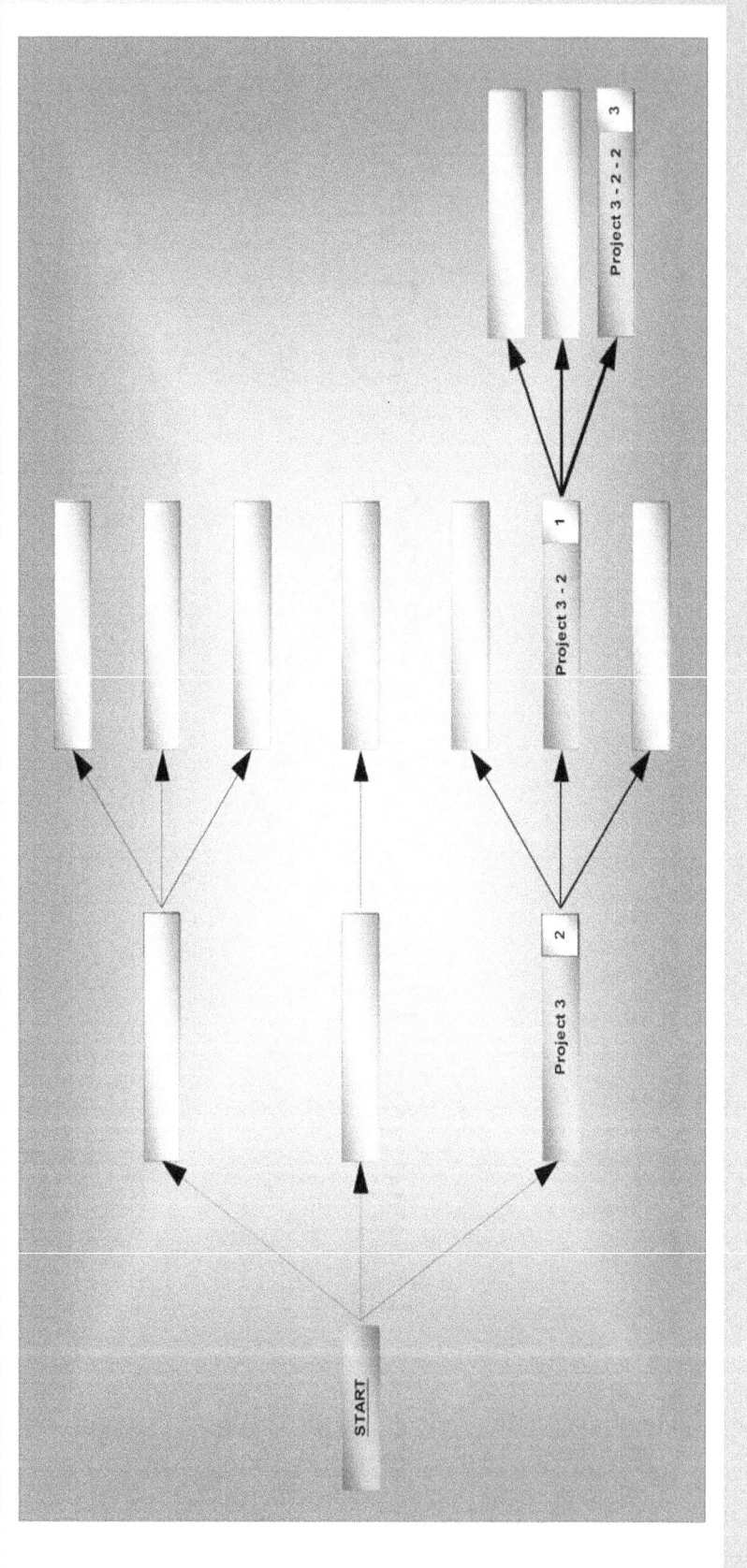

Step 3 – Dynamic Decision Tree

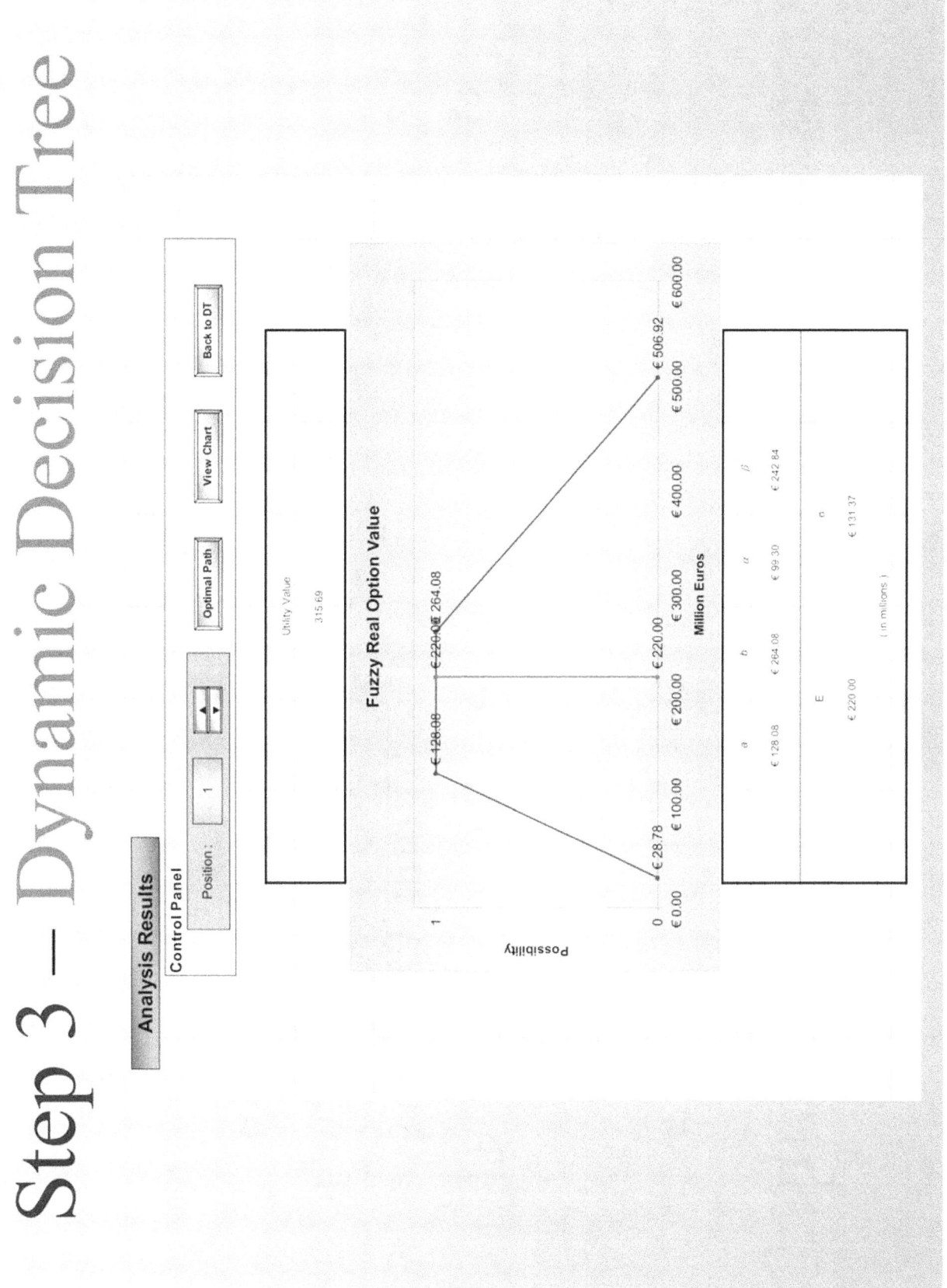

Step 3 – Dynamic Decision Tree

Fuzzy NPV Analysis
Fuzzy Black-Scholes Theory
Dynamic Decision Tree

The output of our methods is not real numbers but <u>fuzzy numbers</u>

Ranking fuzzy numbers is generally impossible

We have employed value function to order fuzzy real option values

Step 3 – Dynamic Decision Tree

Considering fuzzy number of trapezoidal form

$$C = (a, b, \alpha, \beta)$$

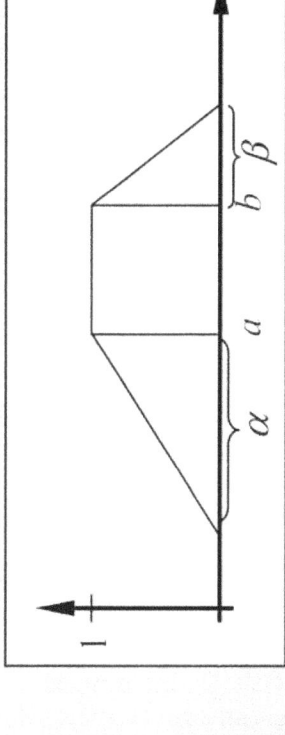

the following utility function has been applied

$$u(C) = \frac{a+b}{2} + r_A \cdot \frac{\beta - \alpha}{6},$$

where $r_A \geq 0$ denotes the degree of the investor's risk aversion

International Real Option Workshop
Turku May 6.-8., 2002

Markku Heikkilä

User Interfaces and Knowledge Presentation in Real Options – an Approach to Make Real Options Operational

Although option-pricing techniques are well known and widely researched, there are certain thresholds for practising managers to benefit from them in corporate investment planning. These thresholds deal with both quantitative and qualitative features of real options, as well as the way real option models and logic is presented to managers. When it comes to user-modes, or actual contents of decision support, model-centred approach can give only starting points for the managers to understand the usability and applicability of real options.

In order to build good support systems for managers responsible for large giga-size investments certain features of user interface design and knowledge presentation should be taken into consideration. In this article I will present a few of these features based on practical case studies made about planning of gigainvestments in several Finnish industrial corporations. I will also present a framework for overcoming various managerial thresholds based on *process-centred modelling* of decision support. This approach utilises *prototyping* in system construction combined to a model of *the manager's interface to real options,* developed to assess the knowledge requirements of investment planning with real options.

In this article real options are understood as tools for managers to deal with large, giga-sized investments with considerable strategic effects. Majority of real options literature so far deals with building models and methods for assessing real options in normative, mathematical way. Application of such an approach gives analytical results, an outcome of the analysis process. The practical needs met in cooperation with corporations suggest, however, a more pragmatic, explorative approach to real option valuation. In most cases it is not sufficient for the managers to know only the outcome of the real option analysis, including real option value, model constraints and mathematical computation. Instead a wider managerial approach requires support for both the outcome of the calculation, process of decision making and learning of real options in management.

Two approaches of knowledge presentation in real options

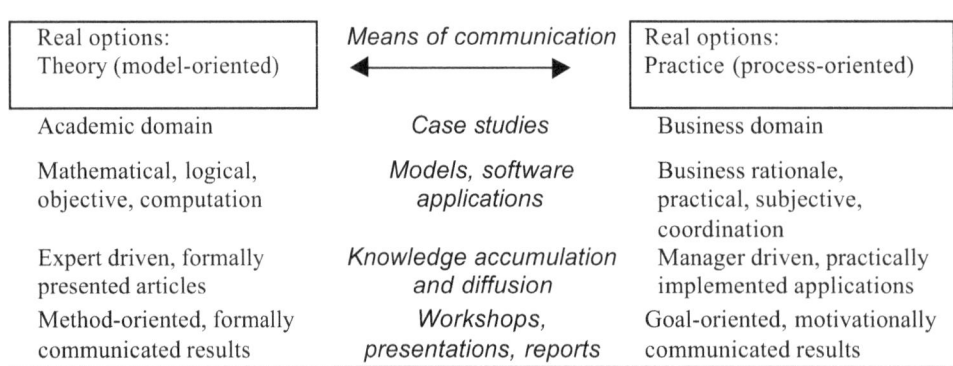

Real options: Theory (model-oriented)	*Means of communication* ◄──────►	Real options: Practice (process-oriented)
Academic domain	*Case studies*	Business domain
Mathematical, logical, objective, computation	*Models, software applications*	Business rationale, practical, subjective, coordination
Expert driven, formally presented articles	*Knowledge accumulation and diffusion*	Manager driven, practically implemented applications
Method-oriented, formally communicated results	*Workshops, presentations, reports*	Goal-oriented, motivationally communicated results

Figure 1 Knowledge presentation in real options

DAY 2

SESSION 4

Marco Guimaráes Dias
"Overview of Real Options in Petroleum"

*Francisco Alcaraz

*Riikka Haasanen
Fortum Case

*No Material

Real Options in Upstream Petroleum: Overview of Models and Applications

By: Marco Antonio Guimarães Dias[1]

This Version: April 27th, 2002

Abstract

Investment in projects of exploration and production (E&P) of petroleum must consider several technical and economic uncertainties which are inherent in this industry. Under these conditions, there are many valuable real options that add value and change the investment decisions when compared with more traditional methods. This article presents an overview of E&P real options models, highlighting the main classical models, including a discussion of the stochastic models for oil prices in real options applications. Some important E&P applications are presented, such as the selection of mutually exclusive alternatives under uncertainty, the wildcat drilling decision, the appraisal investment decisions, and the analysis of option to expand the production through optional wells. This paper also presents a practical way to incorporate technical uncertainties together with economic uncertainty in continuous-time models.

Keywords: real options, investment under uncertainty, E&P investments, project valuation under uncertainty

[1] Technical Consultant by Petrobras and Doctoral Candidate by PUC-Rio. He is the author of the first website on real options at
http://www.puc-rio.br/marco.ind/ Emails: marcoagd@pobox.com and marcodias@petrobras.com.br
This version was prepared to the International Real Option Workshop 2002 - Turku, Finland, May 2002.

1) Introduction

Exploration and production (E&P) of petroleum is an activity rich in real options, which if managed optimally enhance the value of the portfolio of projects and real assets in general for the oil company. The Figure 1 below displays the sequential options process for the typical phases of E&P investment decisions. In the exploration phase managers face the wildcat drilling decision, which in most cases is optional. In case of success (discovery of a reserve), the firm has the option to invest in the appraisal phase through delineation wells and additional 3D seismic, in order to get information about the size and quality of the reserves, reducing the technical uncertainty. When the level of remaining technical uncertainty does not justify additional investment in information, the firm has the option to develop the reserve committing a large investment in development (D) or to give back the undeveloped reserve to the National Petroleum Agency. Finally, the firm has operational options during the producing life of the reserve, that is, has options like the option to expand the production (e.g.: by adding optional wells), option of temporary suspension of the production or even option to abandon the concession.

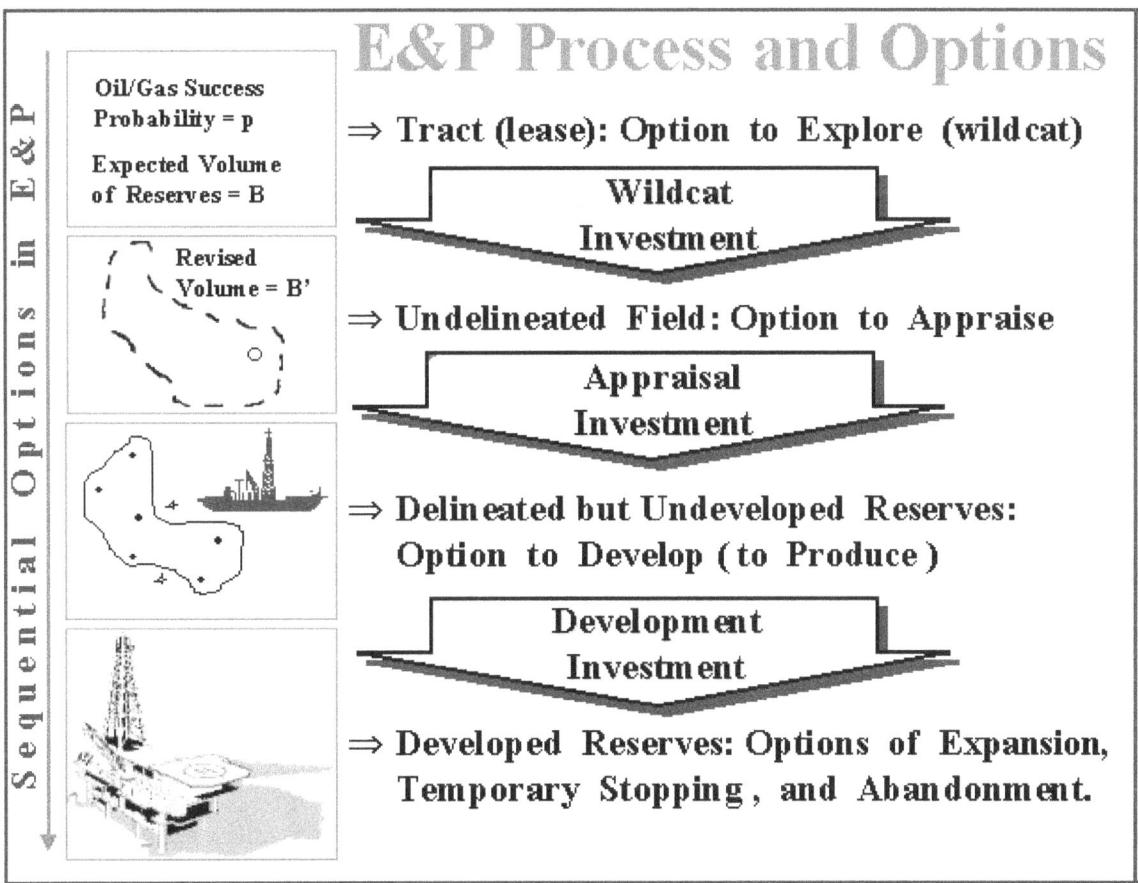

Figure 1 – Exploration & Production as a Sequential Real Options Process

Most of the earlier real options models were developed for natural resources applications, due to the availability of large data for commodities prices. The main earlier models were Tourinho (1979), first to evaluate oil reserves using option pricing techniques, Paddock, Siegel & Smith (1988), a classical model discussed in the next section, Brennan & Schwartz (1985), analyzing interactions of operational options, and Ekern (1988), valuing a marginal satellite oilfield. See Dias (1999) for a bibliographical overview.

A sample of other important real options models for petroleum applications are briefly described in sequence, highlighting the main individual contribution. Bjerksund & Ekern (1990) showed that for initial oilfield development purposes, in general is possible to ignore both temporary stopping and abandonment options *in the presence* of the option to delay the investment. Jacoby & Laughton (1992) showed an application to oilfied development under a complex tax system. Kemma (1993) described some case studies in her long consultant for Shell. Beliossi (1996) presented the valuation of oil companies and managerial efficiency. Schwartz (1997) compared oil prices models that are discussed in the next section. Laughton (1998) found that although oil prospect value increases with both oil price and reserve size uncertainties, the greater oil price uncertainty delays all options exercise (from exploration to abandonment), whereas exploration and delineation occur sooner with reserve size uncertainty. Cortazar & Schwartz (1998) applied the flexible Monte Carlo simulation to evaluate the real option to develop an oilfield. Pindyck (1999) analyzed the long-run behavior of oil prices and the implications for real options. The textbooks of Dixit & Pindyck (1994, mainly chapters 12), Trigeorgis (1996, pp. 356-363) and Amram & Kulatilaka (1999, chapter 12) also analyzes models for investment in oil and natural resources industry.

This paper is organized as follows. In the second section are described the classical real options models applied to E&P and a discussion on the stochastic modelling for oil prices. In the third section is presented an application of selection of alternatives to develop an oilfield under uncertainty with help of the concept of economic quality of a developed reserve. The fourth section analyzes the wildcat drilling problem considering the revelation of information. The fifth section presents a simple way to include technical uncertainty into a dynamic model, to model the investment in information. The sixth section presents the option to expand the production through optional wells. The seventh section concludes the article.

2) Classical Real Options Models in Petroleum and the Oil Price Modelling

In the beginning of 80's, Paddock, Siegel & Smith started a research in the MIT Energy Laboratory using options theory to study the value of an offshore lease and the development investment timing. They wrote a series of papers, two of them published in 1987 and 1988. The Paddock, Siegel & Smith (PSS) approach is the most popular real options model for upstream petroleum applications. This classical model is useful

3

for both learning purposes and as first approximation for investment analysis of development of oil reserves, even thinking that, in real life, are necessary models that fit better the real world features. The book of Dixit & Pindyck (1994, see chapter 12) describes this model in a more compact and didactic way. This model has practical advantages (when compared with others options models) due its simplicity and few parameters estimation. One attractive is the simple analogy between Black-Scholes-Merton financial option and the real option value of an undeveloped reserve, which is illustrated in the Table 1.

Table 1 – Analogy Between Financial Options and Real Options

Black-Scholes-Merton's Financial Options	Paddock, Siegel & Smith's Real Options
Financial Option Value	Real Option Value of an Undeveloped Reserve (F)
Current Stock Price	Current Value of Developed Reserve (V)
Exercise Price of the Option	Investment Cost to Develop the Reserve (D)
Stock Dividend Yield	Cash Flow Net of Depletion as Proportion of V (δ)
Risk-Free Interest Rate	Risk-Free Interest Rate (r)
Stock Volatility	Volatility of Developed Reserve Value (σ)
Time to Expiration of the Option	Time to Expiration of the Investment Rights (τ)

The analogy above is useful for other real options applications. Instead developed reserve value is possible to consider any operating project value (V) as the underlying asset for this option model. In absence of a direct market value for V, is possible to compute V as the revenue net of operational costs and taxes. So, the traditional *net present value* (NPV) is also equal to V – D, where D is the present value of the investment cost to develop the project, and the analogy works in similar way.

The time to expiration of this real option model is the deadline when the investment rights expire. This is named relinquishment requirement, when the oil company faces a "now-or-never" opportunity: or presents and compromises with an immediate development investment plan or return the concession rights back to the National Petroleum Agency. This time varies from 3 to 10 years.

The current value of developed reserve V can be calculated as the present value of the revenues net of operational costs and taxes. But if sufficient data from the market of developed reserves is available, V is just the price of an underground barrel of reserve (v, in dollars per barrel, $/bbl) times the quantity of barrels in the ground, the reserve size B (that is, V = v .B). See data, references and discussion on the market value of a developed reserve in Adelman, Koehn & Silva (1989, data in Table 2). In the next section is presented a model that V is proportional to P, that is, v = q.P or V = q.P.B, where the proportional factor q is named economic quality of the developed reserve.

4

The investment cost to develop the reserve D is the present value of the investment flow necessary to develop the reserve. This is analog to exercise price of the financial option because it is the commitment that the Oil Company faces when exercising the real option to develop the oilfield.

The volatility σ is the annual standard deviation of dV/V. In the case that V is proportional to P is possible to use the same value of the volatility of oil prices[2], which has more available data for estimations. Dixit & Pindyck (1994, chapter 12) recommend σ between 15% and 25% per annum. Some authors (e.g.: Baker, Mayfield & Parsons, 1998, p.119) use a value higher than 30%p.a. for σ. Other possibility is to estimate σ using a Monte Carlo simulation of the stochastic processes for P, operational cost and taxes, using the cash-flows equations, in order to get a combined distribution of the return of V. See the book of Copeland & Antikarov (2001, chapter 9) for a detailed approach of this alternative.

The analog to dividend yield, the cash flow yield δ can be viewed as a net cash flow as a percentage of V. In the case of petroleum reserves, there is a depletion phenomenon due the finite quantity of petroleum in oilfield and a consequent *decline rate* (ω) in the oilfield output flow rate along the life of the reserve. The equation to estimate δ including depletion are presented in Paddock, Siegel & Smith or in Dixit & Pindyck (1994)[3], but two more practical ways to estimate δ are presented below.

For the model that assume the value of a barrel of developed reserve as proportional to the price of a (wellhead) barrel of oil, is possible to see δ as the (net) convenience yield of oil prices by using data from futures market. With the notation $\Phi(t)$ for the oil futures price for delivery at time t, P for the spot oil price (or the earliest futures contract), r for the risk-free interest rate, and Δt = time interval, the equation is:

$$P = \Phi(t) \cdot \exp[-(r-\delta) \cdot \Delta t] \qquad (1)$$

The second way is a practical rule using a long-term perspective that is useful for real options models. What is a good practical value for the net convenience yield δ? Pickles & Smith (1993) suggest the risk-free interest rate. They wrote (pp.20-21): *"We suggest that option valuations use, initially, the "normal" value of net convenience yield, which seems to equal approximately the risk-free nominal interest rate"*[4].

---

[2] If V follows a geometric Brownian motion and V is a constant multiple of the price P (that is V = k P), then P also follows a geometric Brownian motion and with the same parameters (σ, δ, α). See Dixit & Pindyck (1994, p. 178).

[3] Complete proofs for the equations of Paddock, Siegel & Smith is available at www.puc-rio.br/marco.ind/petmodel.html

[4] In the risk-neutral valuation, largely used in options pricing, the risk-neutral drift is r − δ and using the suggestion of Pickles & Smith (r = δ) we get a *driftless* risk-neutral process, which sounds reasonable for the long-run equilibrium.

5

One interesting feature of the PSS model is the resultant *partial differential equation* for the real option value (the value of undeveloped reserve) is identical to Black-Scholes equation with continuous dividend. Only in the boundary conditions appear two additional conditions, to take account for the earlier exercise feature of this American type call option (most real options are modeled as American options rather than European options, to take account of the early exercise feature). See these equation and boundary conditions for example in the book of Dixit & Pindyck (1994, chapter 12). By solving this partial differential equation numerically we obtain two answer, the *real option value* of the undeveloped reserve (F) and the *decision rule* (invest or wait for better conditions). Thanks to the analogy, any good software that solves American call option (with continuous dividend-yield) solves this real options model.

The decision rule is given by the critical value of V (named V*) which the real option is "deep-in-the-money", so that is optimal the immediate exercise of the option to develop the oilfield when V(t) ≥ V*(t). For models where V is proportional to P, it is easier to think with P*, the oil price that makes a particular undeveloped oilfield to be "deep-in-the-money", and the rule is to invest when P ≥ P*.

Figure 2 presents a typical solution for an undeveloped oilfield with 100 million barrels (and q = 0.187, so that V = 18.7 x P). The curves represent the real option values for the cases of five years to expiration of rights (τ = 5 years), one year (τ = 1 year), and at the expiration (τ = 0). This last case, known as "now-or-never" case, the NPV rule holds and the real option value is the maximum between the NPV and zero.

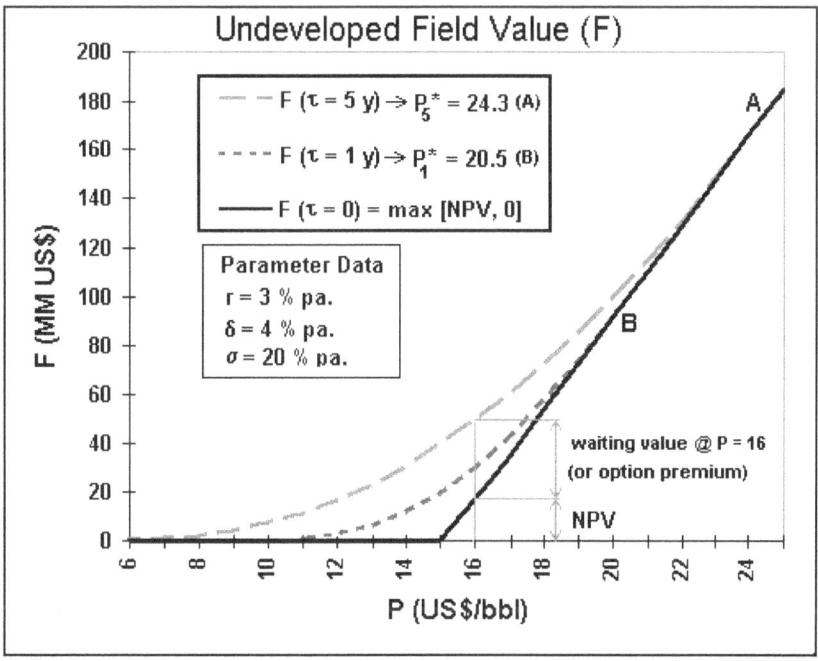

Figure 2 – The Real Option Value of an Undeveloped Reserve

6

In the figure above, for P = 15 $/bbl the NPV is zero (so, 15 $/bbl is the break-even oil price). The figure shows that at 16 $/bbl the NPV is positive, but the net waiting value, also named *option premium* (it is the difference F – NPV), for τ = 5 years is not zero (it is even higher than the NPV). So, at $16/bbl the optimal policy for this field-example is to "wait and see". The option premium is zero only at the tangency point between the option curve and the NPV line (the now-or-never line), which occurs only at the point A, at P*(τ=5) = 24.3 $/bbl. At this point, even with 5 years to expiration, the oilfield is "deep-in-the-money" and is optimal the immediate exercise of the option to invest. In the figure is also displayed the real option curve for one year to expiration (the option is less valuable than the case of 5 years), and the critical value (or threshold value) P* drops to 20.5 $/bbl (see point B). Figure 3 below presents the threshold curve for this oilfield, from 5 years to expiration until the "now-or-never" deadline[5].

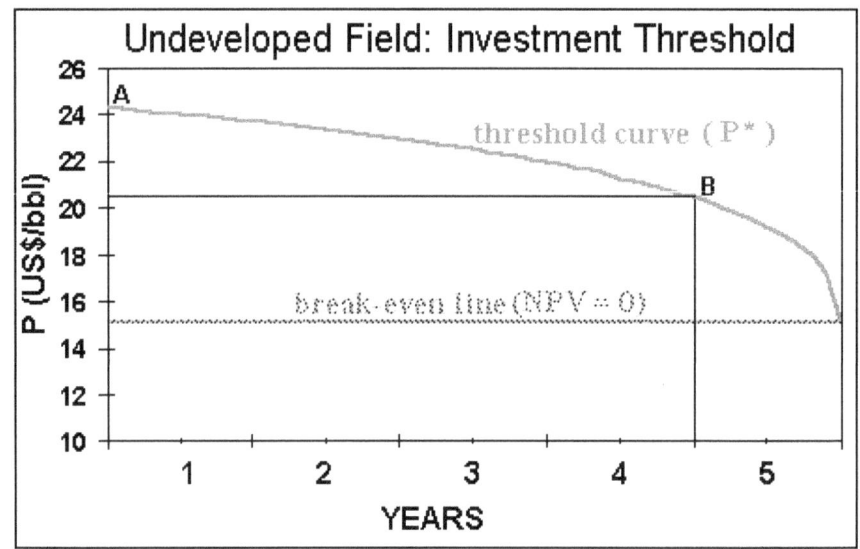

Figure 3 – The Real Option Decision Rule: Invest Above the Threshold

Note in the Figure 3 the correspondence with the points A and B of the Figure 2. Note also that at the expiration (now-or-never), the real option rule collapses to the NPV rule, that is, invest if the oil prices is higher than the break-even price for the oilfield (in this case about 15 $/bbl).

Like Black-Scholes, this model assumes that the underlying asset (here the reserve value V or the oil price P itself) follows a random-walk model named Geometric Brownian Motion (GBM). Under this model, the oil prices (and V) in the future have a lognormal distribution with variance that grows with the forecasted horizon, and a drift that grows (or decays) exponentially. The Figure 4 illustrates this process.

---

[5] This example was solved using the spreadsheet "Timing", a shareware available at the author website.

7

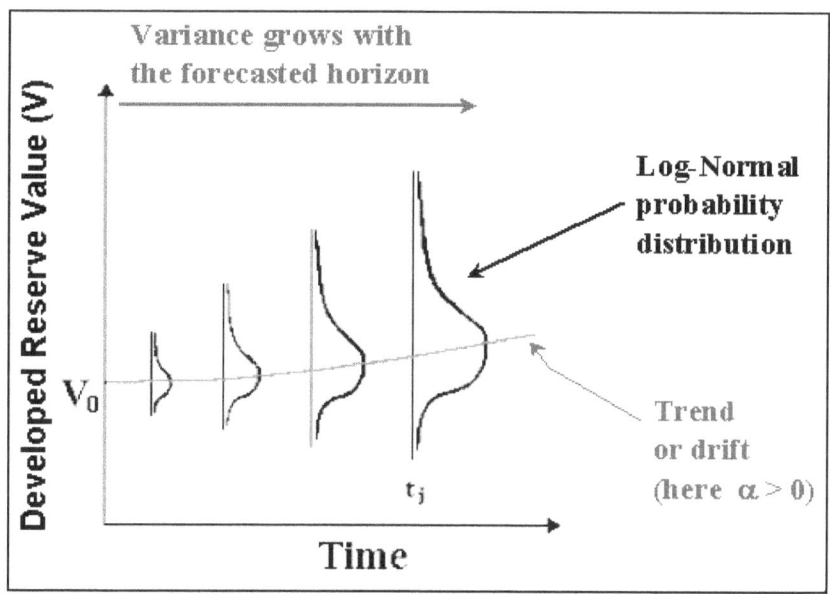

Figure 4 – Map of Probabilities for the Geometric Brownian Motion

But most specialists argue that a stochastic model for commodities must consider the mean-reverting feature. The idea is that if the price is too far (above or below) a certain long-run equilibrium level $\overline{P}$, market forces (including OPEC) will act to reduce (if $P \gg \overline{P}$) or increase (if $P \ll \overline{P}$) the production (for OPEC) and E&P investment (for non-OPEC oil companies). This creates a reverting force that is like a spring force, as strong as P is far from the equilibrium level $\overline{P}$. Figure 5 illustrates the mean-reversion case.

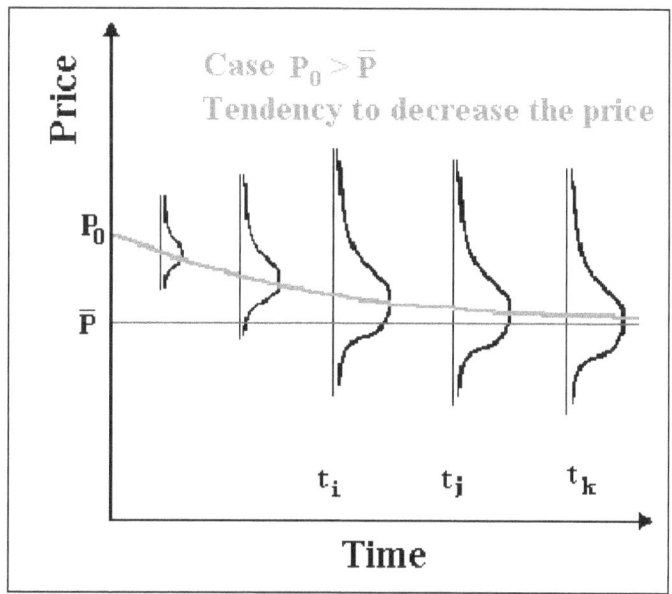

Figure 5 – Map of Probabilities for the Mean-Reversion Model

8

The drift curve (in red) is the curve of expected value for P. For mean-reversion models this curve evolve toward the equilibrium level $\overline{P}$. This drift is consistent with the *backwardation* phenomenon in futures market, when the price of a futures contract is inferior to both spot price and shorter-maturity contracts. Although the distribution of futures prices is also lognormal as in the GBM model, note in the picture that the variance of the distributions grows until a certain time t_i and remains constant after this. The reason is the reversion force effect that does not permit, even in a distant future, to be likely extreme values for P.

The mean-reversion model is more consistent with futures market, with econometric tests and even with microeconomic theory. However the GBM model is much simpler to use. The natural question is: Is wrong to use the GBM model, as in the Paddock, Siegel & Smith, to model oil prices in real options applications? Is significant the error by using GBM instead the mean-reversion model (MRM)? What is a good value for the equilibrium level $\overline{P}$? This paper discuss these and other related questions analyzing the point of view from the two main researchers in commodity prices, Robert Pindyck (MIT) and Eduardo Schwartz (UCLA), in recent articles.

Pindyck (1999) discuss the long-run evolution of the oil prices using 127 years of data and found mean-reversion but the reversion is slower (half-life[6] of 5 years) than presented in some other papers (half-life from one to two years). The oil prices revert to a quadratic U-shaped trend line (instead a fixed level as showed in the Figure 5), which is pointed as consistent with models of exhaustible reserves incorporating technological change. He presents a model for oil prices named *multivariate Ornstein-Uhlenbeck* where the oil prices revert toward a long-run equilibrium level that is itself stochastic and mean-reverting.

In other words, Pindyck argues that the mean-reversion model is better for oil prices, but the equilibrium level is uncertain and oscillates with the time. In addition, his econometric tests show that the reversion process is slow (for high prices case, one reason is that investment in E&P take many years to put new oil in the market). He concludes that for applications like real options '*the GBM assumption is unlikely to lead to large errors in the optimal investment rule*'. This conclusion is reaffirmed in his more recent study (Pindyck, 2001). So, the simple model of Paddock, Siegel & Smith is a not bad approximation.

Schwartz (1997) compares 4 models for the stochastic behavior of oil prices: geometric Brownian motion, pure mean-reversion model, two-factor model and three-factor model. Pure reversion model is the MRM for a fixed (non-stochastic) level and without any other additional stochastic variable or process. The original two-factor model is due Gibson & Schwartz (1990), which oil prices (P) follow a geometric

---

[6] Half-life here is the expected time for the price reaches the half of the distance between P and the equilibrium level.

9

Brownian motion but the convenience yield (δ) of oil follows a mean-reverting process. The three-factor model is like the two-factor but has the interest-rate (r) as third stochastic variable (modelled as mean-reverting).

Schwartz (1997) prefers the two and three factors models, which are less predictable than the pure mean-reverting model. Although in a real options example he shows that the pure reversion model induced higher error (compared with the two and three factors model) than the GBM in the investment decision, he alerts that GBM can induces investment too late because neglects mean-reversion.

Considering both points of view (Pindyck and Schwartz) for real options models, I think that GBM is a good approximation in several cases, but it is not an adequate when the spot prices are too far from a more reasonable long-run equilibrium level (actually this level could be in the range of 18-22 $/bbl). The slope of the term-structure of futures market is a good intuitive way to see if spot price is too far or not.

For the GBM model, every change in oil price is a permanent change in the long-run price drift, whereas the pure mean-reversion assumes the opposite, that is, that every price oscillation is just a temporary deviation from the predictable long-run equilibrium level. Both point of views seems too drastic. A more reasonable point of view is a model not too unpredictable as GBM, neither too predictable as pure mean-reversion. There are three classes of models in this intermediate point of view that is discussed below. The table below summarizes the main available stochastic models for the oil prices.

Table 2 – Stochastic Models for Oil Prices in Real Options Applications

Type of Stochastic Model	Name of the Model	Main Reference
Unpredictable Model	Geometric Brownian Motion (GBM)	Paddock, Siegel & Smith (80's)
Predictable Model	Pure Mean-Reversion Model (MRM)	Schwartz (1997, model 1)
More Realistic Models	Two and Three Factors Model	Gibson & Schwartz (1990), and Schwartz (models 2 and 3)
	Reversion to Uncertain Long-Run Level	Pindyck (1999) and Baker, Mayfield & Parsons (1998)
	Mean-Reversion with Jumps	Dias & Rocha (1998)

As in Pindyck (1999) and in Baker, Mayfield & Parsons (1998), Schwartz & Smith (2000) also present a model of mean-reversion toward an uncertain long-run level. Schwartz & Smith conclude that this model is equivalent to the two-factor model. Interesting, they also conclude that for many long-term investment,

10

we may be able to safely evaluate investments using uncertainty in equilibrium prices only, modeled as GBM!

The third more realistic model presented in the table is due to Dias & Rocha (1998), which consider mean-reversion for oil prices in normal situations[7] but allow for large jumps in the oil prices due abnormal news, during the period before the investment. For the typical time-step of E&P investment (quarters or year), sometimes occurr an abnormal news like war, market crashes or OPEC surprises, that give a radical change of expectations about the balance of supply versus demand, causing large variations in few weeks or months. Figure 6 shows the oil prices (average monthly data) indicating jumps-up and jumps-down.

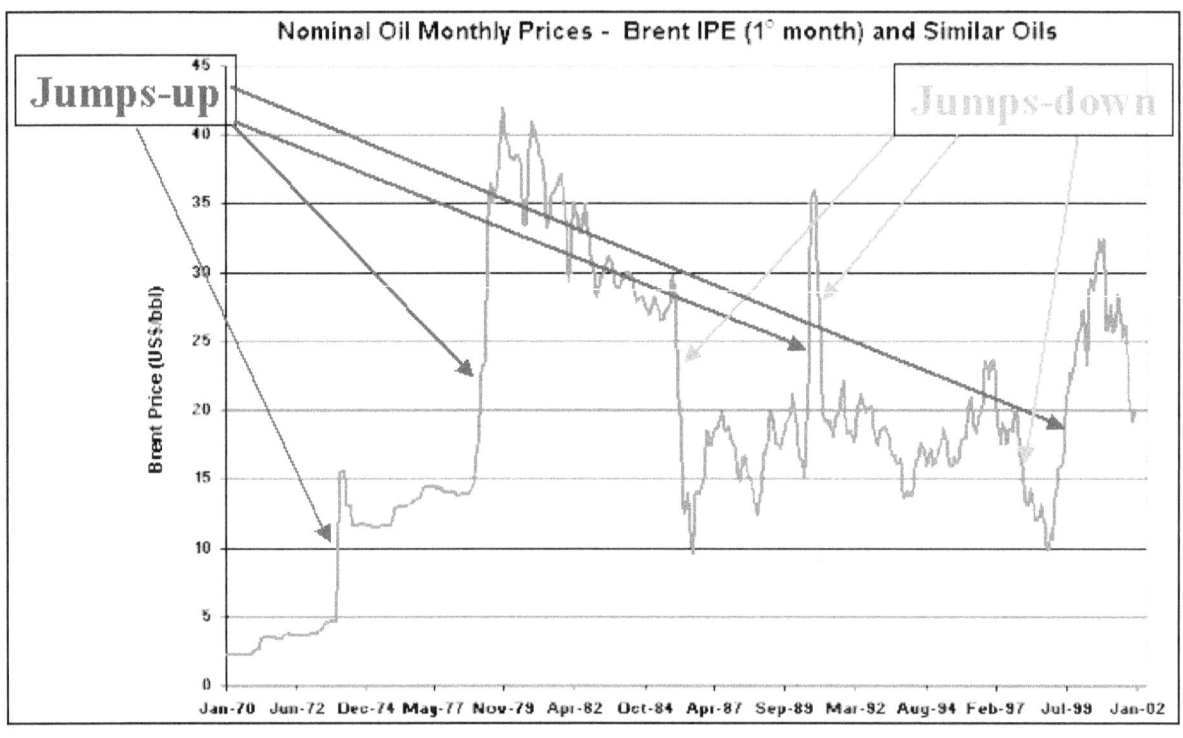

Figure 6 – Oil Prices and Large Jumps in the Last 30 Years

Like the models of two-factor and of reversion to an uncertain equilibrium level, the jump-reversion model is not too predictable as the pure mean-reversion model (due to the jumps component and the linear relation between NPV and P) neither too unpredictable as GBM (due to the reversion component). The model feature of recognizing the possibility of large jumps in some cases can induce better corporate decisions. For example, in oil-linked credit securities and in others oil prices linked agreements, the jumps feature highlights the convenience to put "cap" and/or "floor" in the credit spread or in profit share that is

---

[7] Dias & Rocha (1998) reversion is toward a fixed long-run equilibrium of $20/bbl. However, they suggest to set uncertainty in the equilibrium level (perhaps a GBM with low volatility) as the main improvement for the model.

11

linked to oil prices. In December 1998, the Brent oil prices had dropped to 10 $/bbl. At this time Petrobras and the other side considered a model of mean-reversion with jumps and set *cap* and *floor* in an important 10 years "win-win" contract linked to oil prices. One year latter the oil prices rose about 150% and the cap protected Petrobras to pay more than the desirable due the jumps in the year of 1999. Few months later, the oil prices reached 30 $/bbl, three times the oil prices at the contract date, and the cap protection remained important.

3) Selection of Mutually Exclusive Alternatives and Economic Quality of a Developed Reserve

As mentioned before, a simple model for the *developed reserve value* V is the proportional model that V is a linear function of the price P. The equation is:

$$V = q \cdot P \cdot B \tag{2}$$

Where B is the reserve size (in million of barrels), V is the developed reserve value (in million dollars) and q is the *economic quality of the developed reserve* (see discussion below). Considering the development cost of the reserve D, the equation for the net present value (NPV) of the project is:

$$NPV = V - D = (q \cdot P \cdot B) - D \tag{3}$$

These equations tell that V and NPV has a linear relation with the oil prices (P). Is reasonable this linear assumption? As pointed out by Pickles & Smith (1993, p.16): "*In equilibrium the prices of developed reserves and oil at wellhead must appreciate at the same rate*". Schwartz (1997, his figure 13 and discussion) shows that for a simple cash flow model, either discounted cash flow, GBM, two-factor model and three-factor model, present a linear relation between the NPV and the commodity spot price. Only the pure mean-reversion model presents a non-linear relation.

The linear relation between NPV and P takes real options modelling easier. In the equations 2 and 3, the coefficient "q" is named "economic quality of developed reserve" because for the same oil price (P) and for the same reserve size (B), the market value of the developed reserve V is higher as higher is q. In addition, the product q.P gives the market value of one barrel of developed reserve and it is useful for reserve transactions and merger & acquisitions analysis. The value of q depends of both economic and technical factors. It depends of variables like the discount rate (so depends of the interest rate level and the country risk premium spread), marginal corporate income tax, other taxes, operational costs, reservoir-rock properties (like permeability), fluid quality (quality of the oil and gas), etc.

12

In addition, the economic quality of developed reserve also depends of the amount of capital placed to develop the reserve, mainly the number of wells. A large quantity of wells means faster production and so a higher value for q than a lower number of development wells. However, more wells means higher development cost D, so there is a trade-off between q and D when choosing the number of wells to develop a reserve. By assuming that the reserve size B is approximately the same for all alternatives of development with different number of wells, when changing the development cost D in the NPV equation (3) the only variable that must alter is the quality q. So, the concept of quality q eases the analysis of mutually alternatives to develop an oilfield under uncertainty because the trade-off with investment is captured with a single variable q.

The selection of the best alternative to develop an oilfield for different oil prices is exemplified considering three alternatives, $A_1(D_1, q_1)$, $A_2(D_1, q_1)$, $A_3(D_3, q_3)$, where $D_1 < D_2 < D_3$ and $q_1 < q_2 < q_3$. These alternatives differs due the number of wells and process capacity (more wells means higher production peak). What is the alternative with the higher NPV? The answer depends of the oil price. The Figure 7 below plots the three alternatives NPVs as a function of the oil prices according the linear relation of equation (3). In a real options model this graph corresponds to the "now-or-never" situation.

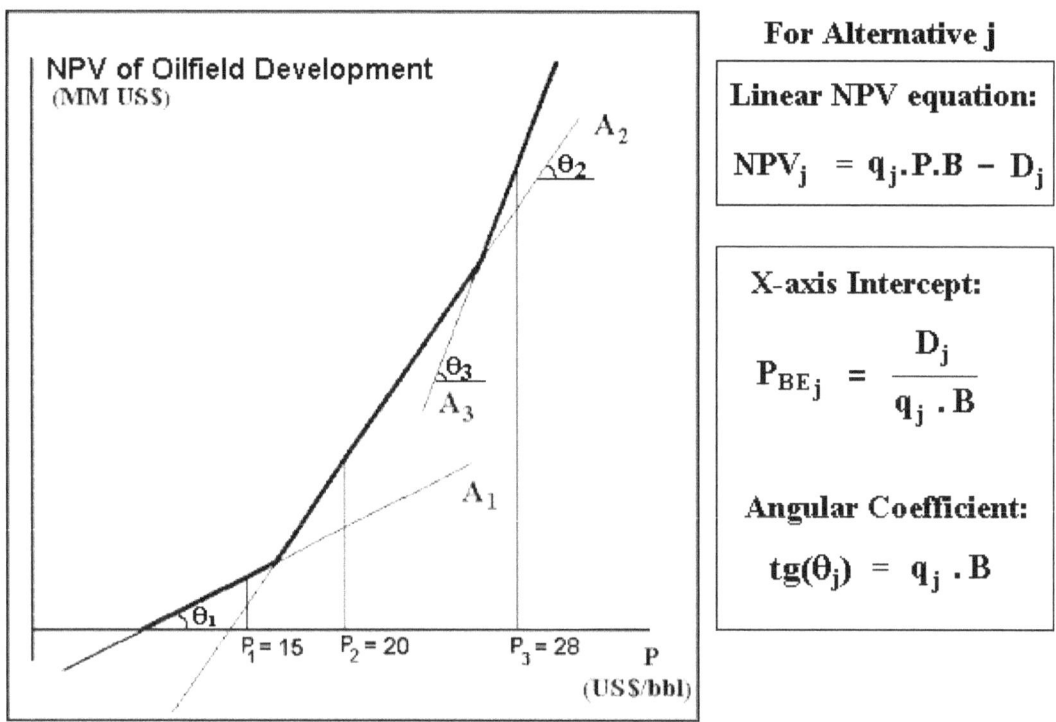

For Alternative j

Linear NPV equation:

$$NPV_j = q_j.P.B - D_j$$

X-axis Intercept:

$$P_{BE_j} = \frac{D_j}{q_j . B}$$

Angular Coefficient:

$$tg(\theta_j) = q_j . B$$

Figure 7 – Three Alternatives: NPVs versus Oil Prices

13

Each alternative has different angular coefficient and different intercepts in the X-axis, and its equations are displayed in the figure above for a general alternative j. The X-axis intercept corresponds to situation of NPV = 0, so the intercept is the break-even price (unitary development cost in $/bbl) for each alternative. While the intercept is related to the development cost, the angular coefficient (tangent of θ_j) is related with the quality q. These parameters are easy to obtain with a discounted cash flow spreadsheet.

Hence, in our example the alternative 3 has an intercept more to the right and higher angular coefficient, due to its higher development cost (D) and higher quality (q). By inspection of the chart above, for low oil prices P_1 the best alternative is the alternative 1, whereas for intermediate price level P_2 the best one is the alternative 2 and so on. So, there are oil price regions where each alternative is better (higher NPV), and the decision depends what region is the oil price.

However, the oil prices are uncertain and in the general case the Oil Company has an option to develop and some time before the expiration of the rights. What is the better decision in this case? Is better to invest if one alternative is deep-in-the money or is better to wait and see for the possibility to invest in a more attractive alternative in a future scenario? The paper of Dixit (1993) is the base of our approach, despite of his model was solved for perpetual real options (for example a land development) and our case is for finite-lived real options. Figure 8 presents the *upper* real options curve for a certain time before the expiration (imagine that the other real options curves are passing under this curve at this time):

14

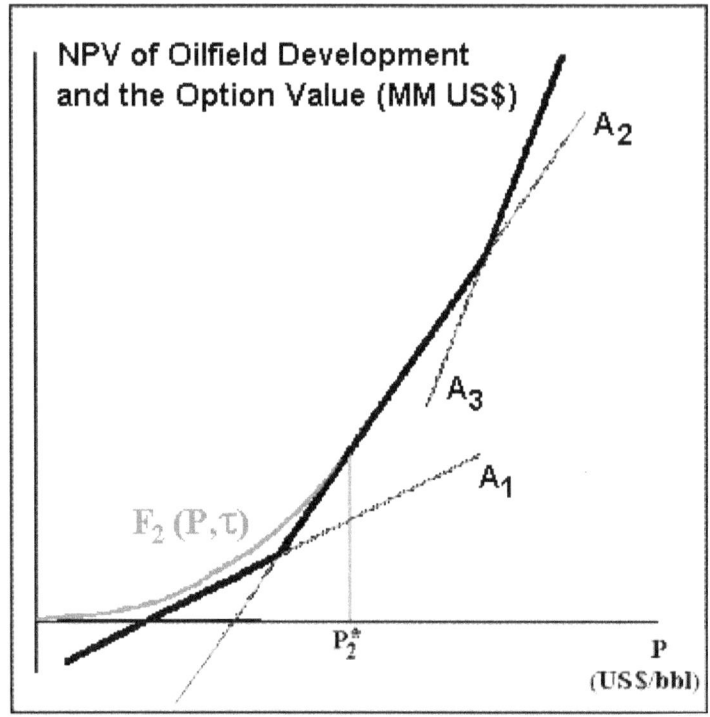

Figure 8 – Selection of Alternatives Before the Expiration

Considering that the option curves for alternatives 1 and 3 at this time pass under the option curve for alternative 2 (showed in the figure above), the decision rule is as follow. If the oil prices at this time is below the critical price P_2*, wait and see even if the alternative 1 seems "deep-in-the-money" because waiting for a scenario which alternative 2 is better, is more valuable (see the chart) than the immediate exercise of alternative 1. However, note that if the alternative 1 quality q_1 increases a little bit, its immediate exercise could be the maximizing value way. Continuing the analysis, if the oil prices is exactly P_2*, the optimal is to develop the oilfield immediately using the alternative 2 (the option to delay is less valuable). Finally, if the oil price is higher than P_2* at this time, the optimal is to exercise the alternative with the higher NPV (alternative 2 or 3, depending of the prices) because the "wait and see" policy is less valuable. By using this approach, is possible to set a complete map of exercise regions along the time, which is one research project that is being developed between PUC-Rio and Petrobras.

4) Investment in Exploration and the Revelation of Information

The linear equation format for the NPV eases the inclusion of technical uncertainties in q and B by simulation together with the economic uncertainty of the oil prices P. Consider the following exploratory example. One oil company (Company X) acquires a tract in a first-sealed bid, with one main prospect and a second marginal prospect in some area of the tract. The tract is placed into a basin with little exploratory

15

historic, but the high oil prices in last two years and acquisitions of tracks in the last three years by several oil companies promise an increasing exploratory activity in this basin. The oil company has 5 years to explore and present a development plan (in case of success). After this expire its rights over the tract.

The marginal prospect today seems unprofitable, a fair estimate of the *expected monetary value*[8] (EMV) is negative. Another oil company (Company Y) offers US\$ 3 millions for the rights of the tract area that contains the marginal prospect. Shall Company X accept the Company Y offers?

The only way to prove the existence of a oil reserve is by drilling a wildcat well. But the drilling cost of this well in deep-waters of this basin cost I_W = 20 MM \$. In addition, the *chance factor* CF (probability of success) is estimate in only 15%. Consider the equation for the expected monetary value (EMV):

$$\textbf{EMV} = -\textbf{I}_W + [\textbf{CF}.\textbf{NPV}] = -\textbf{I}_W + \textbf{CF}.(\textbf{q}.\textbf{P}.\textbf{B} - \textbf{D}) \tag{4}$$

Assume that in case of success, the discovered reserve has an expected size B of 150 millions bbl, and an expected economic quality q of 20%. Suppose that the oil prices today is 20 \$/bbl and is mean reverting. Suppose that the development cost is a linear function of the reserve size B, including a fixed cost (200 MM\$) that represents the minimum development cost to put any production system. The equation for the development cost (MM \$) in this way is given by:

$$\textbf{D} = \textbf{fixed cost} + \textbf{variable cost with reserve size} = \textbf{FC} + (\textbf{VC}.\textbf{B}) = \textbf{200} + \textbf{2}.\textbf{B} \tag{5}$$

Here was assumed that the variable cost of development is 2 \$/bbl. For the expected reserve size B of 150 million barrels, the development cost is \$ 500 million. Using these numerical values into equation (4):

$$EMV = -20 + \{0.15.[(0.2.20.150) - 500]\} = -20 + 15 = \textbf{-5 MM \$}$$

So, the immediate drilling in the prospect has a negative EMV. However, there is 5 years to expiration of the rights and we know that the oil price rises to 22 \$/bbl for example, the EMV become positive. But in addition, there is a lot of technical uncertainty in the estimate of the chance factor (CF), expected reserve size (B) and in the economic quality of the reserve (q), even more due the basin status with few wildcats already drilled there. The large uncertainty in the basin leverages the value of long-term option prospects. In 5 years, many wildcats will be drilled in several tracts of this basin, so that new information about the geology of this basin will be revealed. This *revelation* of geological information shall change the expectations of the variables of the equation (4), because change the geologic and geochemistry models for the basin, the calibration of seismic factors and seismic interpretation, the expectation about the

16

reserve size, etc. The revelation is a free information for long-term prospects like the one under analysis. The revelation can be positive (increasing the EMV) or negative (decreasing the EMV), but both revelations are useful: a negative information can evict a probable misuse of $ 20 million in the wildcat.

In order to ease the evaluation the prospect under these combined technical and market uncertainties, let consider a simpler case that the decision to drill the wildcat in the marginal prospect only will be taken at the expiration in 5 years. The possible values for the EMV at the expiration can be accessed using a *Monte Carlo simulation* of the variables from equation (4). The estimate of the technical uncertainties needs a careful probabilistic analysis of the basin exploratory activity in the next 5 years, and how much this activity can reveal new information. Assuming this job was performed, Table 3 shows the distributions and its parameters for the equation (4) variables, which will be revealed in 5 years. In other words, Table 3 presents the possible scenarios for each variable after 5 years of cumulative new exploratory information.

Table 3 – Distributions for EMV Parameters in 5 Years

Parameter	Distribution	Values
Chance Factor for the Wildcat Well (CF)	Triangular	**Minimum = 0** **Most Likely = 15%** **Maximum = 30%**
Economic Quality for Developed Reserve (q)	Triangular	**Minimum = 15%** **Most Likely = 20%** **Maximum = 25%**
Reserve Size (B) Distribution	Lognormal μ σ	**Mean = 150 MM bbl** Standard-Deviation = 100 MM bbl
Oil Prices (P) Risk-Neutral Distribution	Lognormal μ σ	**Mean = 20 US$/bbl** **Stand.-Dev. = 4.6 $/bbl**

The table also presents the *risk-neutral distribution* for the oil prices, in addition to the technical uncertainty distributions. In the risk-neutral simulation a risk-premium is subtracted from the real drift of the stochastic process (see for example, Trigeorgis, 1996, pp.102 and 218). The use of risk-neutral distribution permits the use of the risk-free interest rate to discount the simulated EMV @ 5 years ahead, in order to get the *present value* of this EMV. The present value of EMV is necessary for oil business

---

[8] Term largelly used in exploratory economics for the value of a prospect considering the chance factor.

analysis like tract rights transactions, in this example to compare with the Company Y buying offer of US\$ 3 million for the tract area that contain the marginal prospect. The other distributions are real distributions because these technical risks have no correlation with market portfolio and do not require risk premium by diversified shareholders of the Oil Company (so the risk-free interest rate is an adequate discount rate). In addition, it is possible to put uncertainty in the costs, for example in the wildcat cost, even more because the drilling rig daily rate are volatile, possibly positively correlated with oil prices.

By combining these uncertainties into the equations (4) and (5) by using a commercial Monte Carlo software, we can get the benefit value (CF . NPV) distribution at t = 5 years, and so the EMV distribution. As we are using as expected value of the distributions the same values used in the today's EMV estimate (so that we are not more optimistic in the future), the expected value of the EMV distribution generated by the Monte Carlo simulation, must be approximately the same (− 5 MM\$). However, the key for valuation of the rights is the optional nature of this wildcat investment. In this future date (5 years ahead) a rational manager only exercise the option to drill the wildcat if the newer estimate EMV is positive. In other words, if the information revealed in the basin exploratory activity combined with the oil prices evolution set a positive EMV for this ex-marginal prospect, the rational manager will exercise the option to drill. In case that the new information reveal that the prospect remains with negative EMV, the option will not be exercised and the prospect value is zero. Real options' thinking introduces an asymmetry in the distribution of the prospect value. The prospect revealed uncertainty combined with the options thinking (optimal exercise of the option) is illustrated in the Figure 9, a "visual equation for the real option value".

18

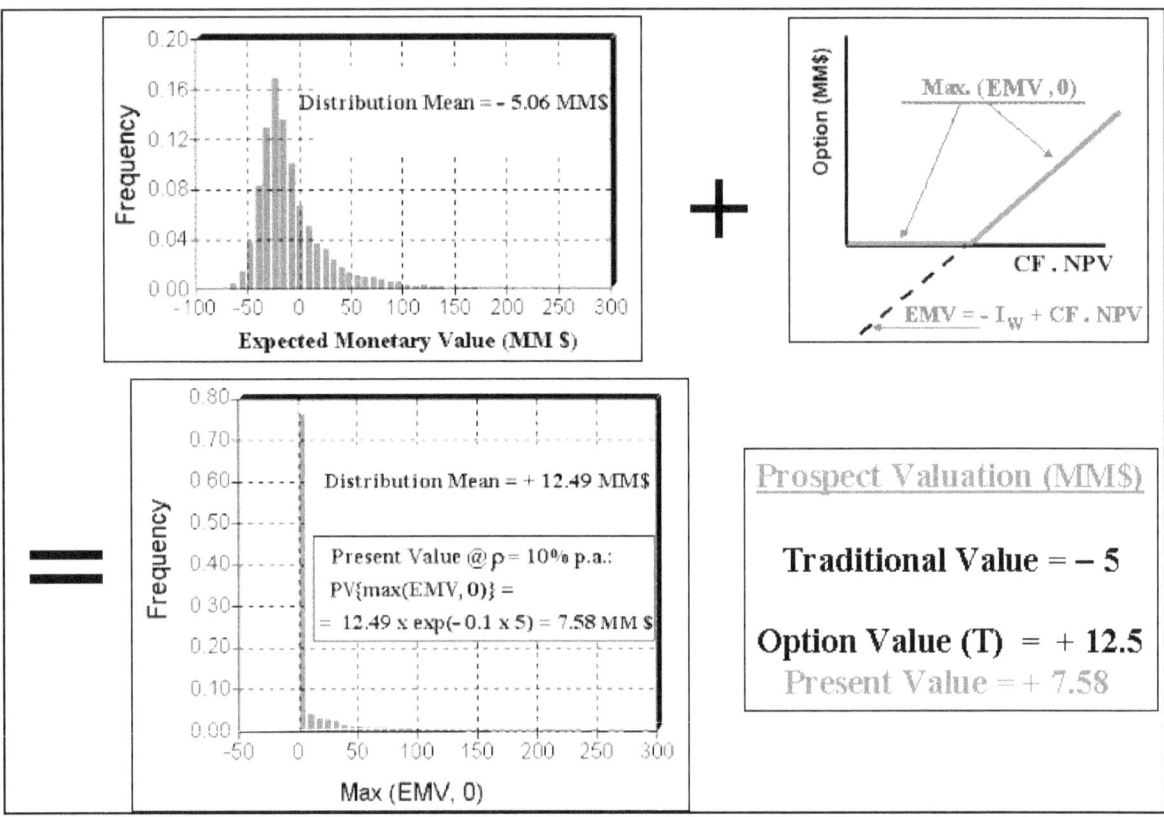

Figure 9 – Visual Equation for Real Options Value of the Prospect

The figure also presents the present value (using a risk-free discount rate of 10% p.a.): the value for the marginal prospect today considering the 5 years to expiration of the real option, is $ 7.58 millions. So, the $3 millions offered by Company Y is too low for this prospect, and Company X must decline it. The value of the prospect in reality is even higher because was not considered the possibility of earlier exercise of the option to drilling, that could be optimal if the prospect become "deep-in-the-money" with sufficient favorable information before 5 years. The real option is of American type, but this European option analysis gives an idea about the prospect minimum value and illustrates the point that real options is leverage by the uncertainty.

5) Alternatives to Invest in Information in a Dynamic Framework

After presenting examples and models for development investment and primary exploratory investment (wildcat drilling), let us present one example for appraisal investment. Suppose that an oilfield is already discovered and even some basic appraisal wells were drilled, but a residual technical uncertainty over the parameters q (reserves quality) and B (reserves size) seems to remain important. Suppose there is two years to expiration of the concession exploratory rights (when the Oil Company must presents an

19

immediate development plan or abandons the concession). Assume that the basin is mature in the sense that the arrival of external exploratory information is not relevant to reveal new expectation over q and B. The only way to get more technical information on the parameters q and B is by investing in information.

There are some alternatives to invest in information, each alternative has different costs and different potential for information revelation. There is even the alternative of not invest in information, named Alternative 0. The investment alternatives are described as follows. Alternative1 is a long-term production test in an already drilled well, with the cost to put a rig over the well to perform the test. Alternative 2 is to drill another appraisal well, a cheaper slim well named ADR (acquisition of data from the reservoir). Alternative 3 anticipates the development drilling schedule, but in a way that optimize the information gathering instead to optimize the drilling costs. See Dias (2001) for a detailed analysis of these cases.

For each alternative we can simulated the scenarios revealed by the new information, and for each scenario revealed we have a normal real options problem that could be solved using models like the described in the section 2 of this article. But this could be computationally very intensive, because for each simulated scenario of q and B (and so D, that is function of B) we have to solve the option problem, finding the threshold curve of P* showed in the Figure 3 of section 2. Is there a more practical way to do this? Is possible to use any kind of threshold curve that is the same of any combination of q, P, B and D? Fortunately the answer is yes for both questions! There is a mathematical artifice that helps us in this job.

The mathematical artifice is a *normalization* of the real options problem that allows a single normalized threshold curve that is the optimal solution for all combinations these parameters. This insight is derived from the method used by McDonald & Siegel (1986, specially p.713) when analyzing a real options problem where both the project value V and the investment cost to develop D follow correlated Geometric Brownian stochastic processes. In order to reduce the two state variable problem to a single state, they observed that a *normalized* threshold (V/D)* is homogeneous of degree zero, in other words, if both V and D are doubled the optimal exercise of real options is the same. As pointed out by Dixit & Pindyck (1994, p.210), the real options value is homogeneous of degree one[9], that is, $F(V, D) = D \cdot F(V/D, D/D) = D \cdot f(V/D)$. Unfortunately, this result that simplifies the combination of technical with market uncertainties in real options problems, was proved only for geometric Brownian motion. Using a Monte Carlo framework, Dias (2001) applies the normalization idea to combine technical uncertainty (that affects both V and D) together with the stochastic process for the oil prices (that affects linearly V).

---

[9] A function is homogeneous of degree n in $\mathbf{x}$ if $F(t\mathbf{x}) = t^n \cdot F(\mathbf{x})$ for all $t > 0$, $n \in \mathbf{Z}$, and $\mathbf{x}$ is the vector of variables.

20

With this normalization, we need to solve the real options problem only one time in order to estimate the normalized threshold curve (V/D)* along the time. With this threshold, we perform Monte Carlo simulations for the technical uncertainties revealed by the investment in information[10], and for the (risk-neutral) stochastic process for the oil prices. Assume that is optimal to invest immediately in information and it is revealed at the *revelation time*. Suppose that the current oil prices is 20 $/bbl, and that in the simulation path, the sampled values were q = 0.2 and B = 100. By using the equations (2) and (5), we get V = 400 and D = 400, so that V/D = 1 at the time t = 0. After this time, considering that no new technical information will be revealed, the value of V/D for this project will oscillate only due the oil prices evolution. For each instant t, we compare the actual V/D with the normalized threshold for optimal exercise of the real options, previously computed. Figure 10 illustrates this procedure, showing two paths, one with exercise and other reaching the expiration date without crossing the normalized threshold curve.

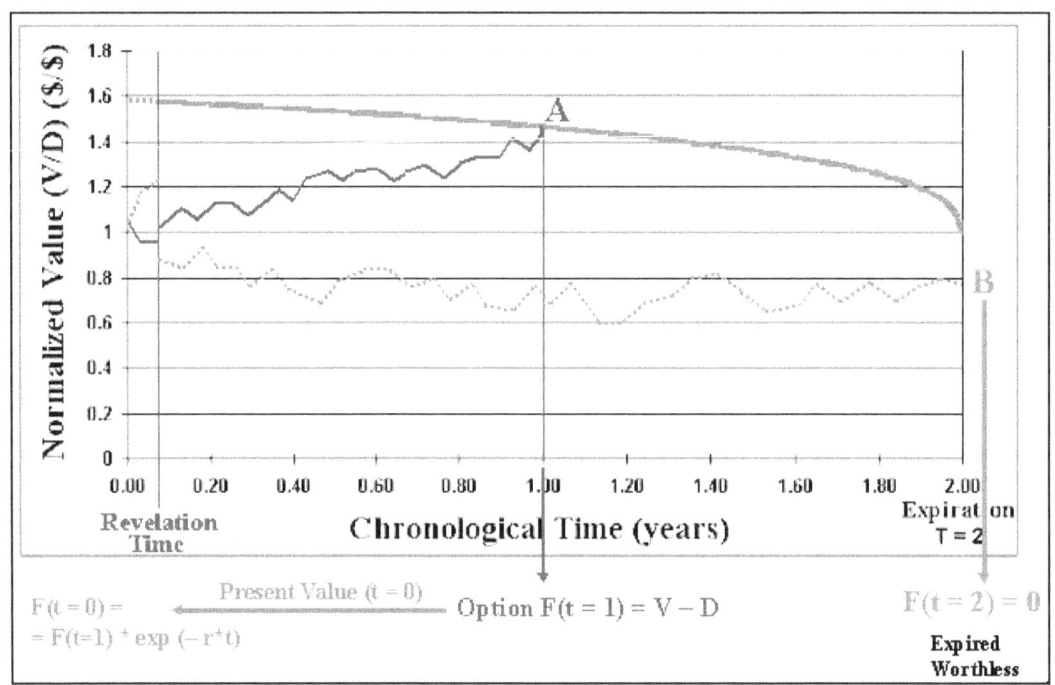

Figure 10 – The Normalized Real Options Approach to Include Technical Uncertainty

If V/D ≥ (V/D)*, we exercise the option (see the figure the point A) and calculate the present value (using the risk-free interest rate). In the opposite case, we wait and see. The paths that do not cross the normalized threshold curve along the two years of option are worthless. By summing all the present values for the real option F(t=0) and dividing for the total number of simulations, we get the real options

---

[10] See Dias (2002) for an in-depth analysis of the distribution of conditional expectations (*revelation distribution*) on the technical parameter after an investment in information.

value after the investment in information. Subtracting the cost of information, we get the value of the final real options value including the investment in information.

Remember that for each alternative, there are different information costs and different distributions for q and B, inclusive the alternative zero, of not invest in information (with information cost equal zero and with single values for q and B instead distributions). This procedure is performed for all the alternatives of investment in information and the real options values considering the information cost for each alternative are compared. The higher one is the best alternative. In general, large uncertainties in q and B will enhance the value for alternatives with higher information revelation potential. Think with the "visual equation for real options" to see this. This simple model above captures this idea.

The last point in this section is the equation necessary to perform the risk-neutral Monte Carlo simulation of the oil price. For the Geometric Brownian Motion, the risk-neutral sample paths are simulated with the equation below, applied to each instant t using the oil price from the previous instant $t-1$:

$$ P_t = P_{t-1} \exp\{ (r - \delta - 0.5\, \sigma^2)\Delta t + \sigma\, N(0,\, 1)\, \sqrt{\Delta t}\, \} \tag{6} $$

In the equation above, Δt represents the time-step in the simulation, $N(0,\, 1)$ is the standard Normal distribution (mean = zero and variance = 1), and the remaining variables are as before. By sampling values from the $N(0,\, 1)$ distribution, we get a whole path along the time and repeating this procedure several times we get several paths. For the real stochastic process (instead the risk-neutral one), just substitute the risk-neutral drift $r - \delta$ by real drift α. But for the real options valuation in complete markets, the risk-neutral one is all that matters for most applications.

6) The Option to Expand the Production Using Optional Wells

Sometimes the best way to reveal information about the remaining technical uncertainties in the oilfield is by using the information generated by the cumulative production in the field and/or measuring the bottom-hole pressure after months or few years of cumulative production. In these cases, the investment in information before the oilfield development is not adequate due the low potential to reveal information in relation to its cost to obtain.

The proper way in this case to get economically the information is to embed options to expand the production into the selected alternative, so that depending of the revealed information, we exercise or not the option to expand the production. In case of oilfield development, a good way is to select some *optional*

22

wells, which will be drilled only in favorable scenarios. In addition, the exactly optimal locations for these wells also depend of the information generated by the initial oilfield production.

There are some costs to embed this option to expand. For example, the processing plant at higher cost could be dimensioned considering this possibility. Other way is by leaving space in the *production unit* (for example an offshore *floating platform*, where the area has a relatively high cost) in order to amply the capacity in case of exercise of the options to drill these optional wells. The third way is to wait the main production decline (reservoir depletion) in order to integrate these wells when the processing capacity permits (see Ekern, 1988). But even in this last case, is necessary to consider the possible new load in the platform from the optional new lines and a possible additional cost because the subsea layout must consider the possibility of optional well to be drilled and its flowlines going to the production platform. However, these costs to embed the optional wells are, in general, only a fraction of the potential benefit of drilling these wells in the favorable scenarios (including favorable market scenarios for oil prices).

The economic analysis of the option to expand with optional wells requires an in-depth study of marginal contribution from each well in the overall oilfield development. In this study is necessary to identify the candidate wells that can become optional wells. Wells with higher reservoir risk are primary candidates because with new reservoir information these wells can become unnecessary or its optimal location could be different. The other class of optional well candidates is that resultant of the marginal well contribution for wells that presented negative NPV or even a small positive NPV[11] so that the option to drill these wells are not "deep-in-the-money".

One feature that in general can be important is the *secondary depletion* in the area of the optional well. The production of the main wells causes a pressure differential in the reservoir with some migration of the oil in the optional well area. In most cases the oil migration doesn't mean that the main wells will produce this oil, or if this oil is produced, it takes too long time to be produced by distant wells. This secondary depletion acts as a dividend lost by the option holder and so has the same effect of dividend yield in the traditional real options models (higher incentive to drill the optional well as higher is the secondary depletion).

The general method to analyze the option of expansion through optional wells is:

- Define the quantity of wells "deep-in-the-money" to start the basic investment in development by the marginal analysis of each well using several reservoir simulations;

---

[11] The direct method to estimate the marginal NPV of a well is by reservoir simulation with and without each well. With the production profiles in each case, the two NPVs are calculated and the marginal NPV is just the difference between these NPVs.

23

- Define the maximum number of optional wells;

- Define the timing (or the accumulated production) that the reservoir information will be revealed;

- Define the of marginal production of each optional well as function of uncertain parameters and the possible scenarios (distribution) to be revealed by the cumulative production;

- Consider the secondary depletion in the optional well area if we wait after learn about reservoir;

- Simplify by limiting the expiration of the option to drill the optional wells (remember the declining NPV due the secondary depletion);

- Add market uncertainty, simulating a stochastic process for oil prices and perhaps for the daily rate of rigs that could drill the optional wells;

- Combine uncertainties using Monte Carlo simulation;

- Use an optimization method to consider the earlier exercise of the option to drill the wells, and calculate option value; and

- Compare this option value with the cost to embed the option to expand. The option value must be higher than the cost to embed the option, to justify this flexibility cost into the development plan.

The Figure 11 below presents the timeline of the option to expand problem that help us understand a practical modelling of this problem.

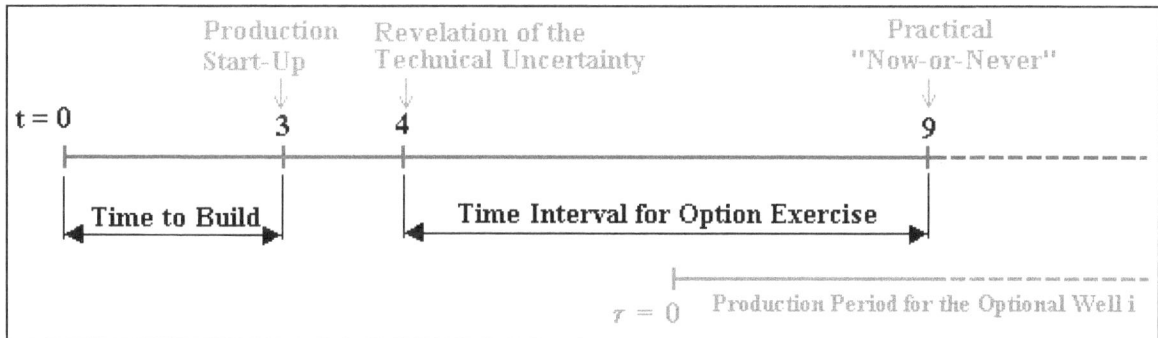

Figure 11 – Timeline for the Option to Expand Model Using Optional Wells

In the figure is considered the time to build of three years to construct the production platform, including the processing plant, the pipelines, the drilling of development wells, etc. In the year 3 occurs the production start-up, and between the years 3 and 4, the initial production of the wells and eventual tests give us all relevant technical information for this example. The period from year 4 until the year 9, the exercise of the option to drill is considered for the optional wells that had reservoir uncertainty before the initial production. The deadline at year 9 is only a practical setting due both the secondary depletion effect and the distant benefit meaning low present value, so that the option value is too small after year 9. The figure also present the timeline for the optional well i that could be drilled at a random instant between years 4 and 9. In this example, the Monte Carlo simulation for the technical uncertainty is placed only at

24

t = 4 and the simulation for the stochastic process for the oil prices start in t = 0 and is necessary only until the exercise of the drilling option (at τ = 0). After the option exercise is necessary only the expected value curve for the oil prices because no other options (like abandonment) are considered that could introduce asymmetries in the value distribution.

The valuation of the option to expand is another research project under development between PUC-Rio and Petrobras. The implementation of this approach needs an active management in all phases of the project. The communication between the project teams in different project phases is necessary to preserve the embedded flexibility of the production unit and subsea layout, against the desire of "optimization" and "cost reduction" that could destroy the option value by not informed project teams in subsequent phases.

7) Concluding Remarks

In this article was described the classical model of Paddock, Siegel & Smith that exploits a simple analogy between American call options and real options model for oilfield development. In addition were discussed different models of the stochastic processes for oil prices, mainly using insights from recent papers of Pindyck and Schwartz, perhaps the two main experts in this subject.

Along this article were illustrated the main cases of the real options process that occur in upstream petroleum applications, showed in the Figure 1. The cases presented were the selection of mutually exclusive alternatives to develop an oilfield; the option to drill a wildcat in an unexplored basin considering the information revelation issue; the option to invest in information in the ending of the appraisal phase, including a mathematical artifice to incorporate the technical uncertainty into a dynamic framework together with oil price uncertainty; and the case of the option to expand using the concept of optional wells.

Several other real options applications of interest for exploration and production of petroleum were not analyzed in this article, for brevity. For example, models for valuation of exploratory prospect including the option to abandon the sequential investment plan in the appraisal phase (see Dias, 1997, for a simple model). The valuation of petroleum reserves with models like Brennan & Schwartz (1985) and Dixit & Pindyck (1994, chapter 7), as performed by Oliveira (1990), is a useful insight mainly for mature reserves, where the options of temporary stopping and abandonment are very important. Other interesting model could be the valuation of deep-waters technology using real options approach. This technology not only adds value to the current portfolio of concessions of an oil company, as also permits that the firm entry in

25

areas and obtain business opportunities that are not available for oil companies that don't have this know-how. The description and analysis of these models are left for a future article.

8) Bibliography

Adelman, M.A. & M.F. Koehn & H. de Silva (1989): "The Valuation of Oil Reserves"
SPE Hydrocarbon Economics and Evaluation Symposium, Proceedings pp.45-52, Dallas, Texas, March 1989 (SPE paper nº 18906)

Amram, M. & N. Kulatilaka (1999): "Real Options – Managing Strategic Investment in an Uncertain World"
Harvard Business School Press, 1999, 246 pp.

Baker, M.P. & E.S. Mayfield & J.E. Parsons (1998): "Alternative Models of Uncertain Commodity Prices for Use with Modern Asset Pricing"
Energy Journal, vol.19, nº 1, January 1998, pp.115-148

Beliossi, G. (1996): "Option Pricing of an Oil Company"
London Business School and Universitá di Bologna. Presented at the 1996 FMA Conference, New Orleans, USA (available at www.puc-rio.br/marco.ind/contrib1.html#Beliossi)

Bjerksund, P. & S. Ekern (1990): "Managing Investment Opportunities under Price Uncertainty: from Last Chance to Wait and See Strategies"
Financial Management, vol. 19, nº 3, Autumn 1990, pp. 65-83

Brennan, M.J. & E.S. Schwartz (1985): "Evaluating Natural Resource Investment"
Journal of Business, vol.58, nº 2, 1985, pp.135-157

Copeland, T. & V. Antikarov (2001): "Real Options – A Practitioner's Guide"
Texere LLC Publishing, 2001, 372 pp.

Cortazar, G. & E.S. Schwartz (1998): "Monte Carlo Evaluation Model of an Undeveloped Oil Field"
Journal of Energy Finance & Development, vol.3, nº 1, pp.73-84

Dias, M.A.G. (2002): "Investment in Information in Petroleum: Real Options and Revelation"
Working Paper, Dept. of Industrial Engineering, PUC-Rio, April 2002, 38 pp.

Dias, M.A.G. (2001): "Selection of Alternatives of Investment in Information for Oilfield Development Using Evolutionary Real Options Approach "
Paper presented at the 5th Annual International Conference on Real Options, UCLA, Los Angeles, July 2001, 29 pp.

Dias, M.A.G. (1999): "A Note on Bibliographical Evolution of Real Options"
in Lenos Trigeorgis, Eds., *Real Options and Business Strategy – Applications to Decision Making* – Risk Books, 1999, pp. 357-362

Dias, M.A.G. (1997): "The Timing of Investment in E&P: Uncertainty, Irreversibility, Learning, and Strategic Consideration"
SPE paper nº 37949, presented at 1997 SPE Hydrocarbon Economics and Evaluation Symposium, Dallas 16-18 March 1997, Proceedings pp.135-148

26

Dias, M.A.G. & K.M.C. Rocha (1998): "Petroleum Concessions with Extendible Options Using Mean Reversion with Jumps to Model Oil Prices"
    Working Paper, first presented at the present at "Workshop on Real Options", Stavanger, Norway, May 1998. Revised version presented in the 3rd Annual International Conference on Real Options, June 1999, Netherlands.

Dixit, A.K. (1993): "Choosing Among Alternative Discrete Investment Projects Under Uncertainty"
    Economic Letters, vol.41, 1993, pp.265-288

Dixit, A.K. & R.S. Pindyck (1994): "Investment under Uncertainty"
    Princeton University Press, Princeton, N.J., 1994, 468 pp.

Ekern, S. (1988): "An Option Pricing Approach to Evaluating Petroleum Projects"
    Energy Economics, April 1988, pp.91-99

Gibson, R. & E. Schwartz (1990): "Stochastic Convenience Yield and the Pricing of Oil Contingent Claims"
    Journal of Finance, vol.45, n° 3, July 1990, pp.959-976

Jacoby, H.D. & D.G.Laughton (1992): "Project Evaluation: A Pratical Asset Pricing Model"
    Energy Journal, vol.13, n°2, 1992, pgs.19-47

Kemna, A.G.Z. (1993): "Case Studies on Real Options"
    Financial Management, Autumn 1993, pp.259-270

Laughton, D.G. (1998): "The Management of Flexibility in the Upstream Petroleum Industry"
    Energy Journal, vol.19, n° 1, January 1998, pp.83-114

McDonald, R. & D. Siegel (1986): "The Value of Waiting to Invest"
    Quarterly Journal of Economics, November 1986, pp.707-727

Oliveira, C.A.P. (1990): "Avaliação e Gerência de Jazidas de Petróleo - Uma Abordagem pela Teoria das Opções" (*Valuation and Management of Petroleum Deposits - An Options Theory Approach*)
    Dept. of Industrial Engineering, PUC-RJ, M.Sc. Dissertation, 1990 (in Portuguese)

Paddock, J.L. & D. R. Siegel & J. L. Smith (1988): "Option Valuation of Claims on Real Assets: The Case of Offshore Petroleum Leases"
    Quarterly Journal of Economics, August 1988, pp.479-508

Pickles, E. & J.L.Smith (1993): "Petroleum Property Evaluation: A Binomial Lattice Implementation of Option Pricing Theory"
    Energy Journal, vol.14, n°2, 1993, pp.1-26

Pindyck, R.S. (1999): "The Long-Run Evolution of Energy Prices"
    Energy Journal, vol.20, n° 2, 1999, pp. 1-27

Pindyck, R.S. (2001): "The Dynamics of Commodity Spot and Futures Markets: A Primer"
    Working Paper, CEEPR, MIT, May 2001, 38 pp.

Schwartz, E.S. (1997): "The Stochastic Behavior of Commodity Prices: Implications for Valuation and Hedging"
    Journal of Finance, vol.52, n° 3, July 1997, pp.923-973

27

Schwartz, E. & J.E. Smith (2000): "Short-Term Variations and Long-Term Dynamics in Commodity Prices"

Management Science, vol.46, nº 7, July 2000, pp.893-911

Tourinho, O.A.F. (1979): "The Valuation of Reserves of Natural Resources: An Option Pricing Approach"

University of California, Berkeley, PhD Dissertation, November 1979, 103 pp.

Trigeorgis, L. (1996): "Real Options - Managerial Flexibility and Strategy in Resource Allocation"

MIT Press, Cambridge, MA, 1996, 427 pp.

28

DAY 3

Wednesday 08.05.

Chairman   Robert Fullér

SESSION 1

09.00 – 10.00 Keynote Speaker

Luis Alvarez, TUKKK, Finland
"Optimal Adaptation of Uncertain Technologies"

10.00 – 10.30 Anett Mehler–Bicher

"Analysis Template for Assessing E-Business Investments by
Applying Real Options"

10.30 – 10.45 Morning coffee

SESSION 2

10.45 – 11.15 Matts Rosenberg

"Capital Investment and Labor Adjustment under Uncertainty
– Empirical Evidence from Finland"

11.15 – 11.45 Christer Carlsson
"Industry Foresight"

11.45 – 12.15 Svante Olofsson
"Agent Applications"

12.15 – 13.15 Lunch

SESSION 3

13.15 – 13.45 Mikael Collan
"A Method for Including Foresight Information into Fuzzy Real Option Valuation"

13.45 – 14.15 Shuhua Liu
"Intelligent Agents"

14.15 – 15.00 EUNITE IBA E and Discussion of Future Research Directions

15.00   Closing of the workshop

DAY 3

SESSION 1

Luis Alvarez
"Optimal Adaptation of Uncertain Technologies"

Anett Mehler– Bicher
"Analysis Template for Assessing E–Business
Investments by Applying Real Options"

Strategic Adoption of Intermediate Technologies: A Real Options Approach

Luis H. R. Alvarez[*] Rune Stenbacka[†]

Preliminary version

April 16, 2002

Abstract

We apply a real options approach to analytically characterize the option value of adopting an intermediate technology. We design an asymmetric duopoly model to delineate how the optimal adoption timing of an intermediate technology depends on the embedded upgrading options available to the firm itself and to its future rival. Focusing on diffusions we develop explicit representations demonstrating that the threshold of adopting an intermediate technology depends negatively (positively) on the leader's (follower's) upgrading intensity. For geometric Brownian motion we explicitly characterize the iso-incentive curves keeping the leader's incentives of adopting the intermediate technology invariant.

Keywords: Real options, strategic technology adoption, intermediate technologies.

AMS Subject Classification: 90A09, 90A11, 62L15, 60G40, 60J60.

JEL Subject Classification: O32 , G30 , D92 , C61.

---

[*]Department of Economics, Quantitative Methods in Management, Turku School of Economics and Business Administration, FIN-20500 Turku, Finland, e-mail: luis.alvarez@tukkk.fi

[†]Swedish School of Economics and Business Administration, Department of Economics, P.O. Box 479, FIN-00101 Helsinki, Finland, e-mail: rune.stenbacka@shh.fi

1

1 Introduction

Recently we have witnessed that European telecommunication operators have made enormous irreversible investments into the mobile UMTS technology. For example, Table 1 shows the price per capita raised for rather similar blocks of spectrum frequencies sold for third generation mobile-phone services in six European countries during year 2000. These spectrum auctions cumulatively raised approximatively US\$ 100 billion, which constitutes more than 1.5 per cent of the combined GDP of these countries. Still the long run forecasts of many experts seem to cast serious doubts on the technological capability as well as the competitiveness of the UMTS technology for the estimated large-scale future demand of datatransmission and internet-based communication. If these perspectives are correct the UMTS technology might represent a fairly temporarily limited phase in the dynamic development of the telecommunications industry. But, why then did the firms invest so heavily in this technology by, for example, offering the spectacular bids we observed in countries like Germany or UK? This study will elaborate the hypothesis that intermediate technologies can be seen as strategic real options, which might play a significantly important role in the competition taking place within the framework of future generations of improved technologies.

Country	Euros per capita
Austria	100
Germany	615
Italy	240
Netherlands	170
Switzerland	20
UK	630

Table 1 Revenues from European 3 G Mobile Spectrum Auctions Completed in 2000.

The adoption of new technologies typically represents a critical component of firms' investment policies. Adoption decisions tend to be predominantly irreversible and the timing of acquiring a new technology constitutes a key strategic decision for a firm. Premature adoption might often be very risky, but could potentially yield significant comparative advantages. In particular, adoption of a currently available, but still intermediate, technology may generate experience and knowledge necessary for improved access to subsequent generations with upgraded versions of the new technology. In line with the influential investment approach manifested in the extensive books by Dixit and Pindyck (1994) and Trigeorgis (1996) the opportunity of adopting a new uncertain technology can be viewed as representing a real option to the firm. In light of the prospect of successive generations of technologies building on each other, the investment outlays of adoption can be cast as a sequence of embedded options.

A decision to adopt a presently available, possibly intermediate, version of a new technology will have ramifications on the firm's future options. In particular, the value to the firm of upgrading to a new generation of technology can be expressed as that of an option to exchange one risky asset, the incumbent technology, for another risky asset, an upgraded technology with a stochastic arrival time in the future. In Alvarez and Stenbacka (2001) we develop an approach based on the Green representation of Markovian functionals for finding the optimal exercise thresholds both of the ordinary real option associated with the updating decision and of the compound real option associated with the incumbent, possibly intermediate, technology. That study also characterizes how the real option values depend on the underlying market uncertainty and on the uncorrelated technological uncertainty regarding future technology improvements. However, the focus of that study is restricted to industries in which there is no strategic interaction between the adoption timings of the firms.

In the present analysis we apply a real options approach based on Green functions to analytically characterize the option value associated with the adoption of an intermediate technology. We formally design an asymmetric duopoly model in order to delineate how the optimal adoption

2

timing of an intermediate technology for a firm (the leader) depends on the embedded upgrading options available to the firm itself as well as to its rival (the follower). The option value is shown to be increasing as a function of the attractiveness of the embedded upgrading options available to the leader, whereas it is decreasing as a function of the quality of the embedded upgrading options which spill over to the follower. In particular, with our focus on stochastic processes evolving according to a geometric Brownian motion and well-specified restrictions on the type of cash flows, we explicitly characterize the optimal timing of when to adopt an intermediate technology. From this characterization we are able to conclude that the threshold of adopting the intermediate technology depends negatively (positively) on the leader's (follower's) upgrading intensity. Furthermore, we explicitly characterize the iso-incentive curves describing the locus of those upgrading intensities (for the leader and the follower) which keep the leader's incentives of adopting the intermediate technology invariant.

Our model can be seen to make a contribution to the analysis of optimal sequential investment behavior for firms facing multi-stage projects. This literature includes Dixit and Pindyck (1994), Dutta (1997), Bhattacharya et al. (1986) as well as Alvarez and Stenbacka (2001). However, these contributions all focus on industries with no strategic interactions between the firms. Relative to the existing literature the value added of the present analysis lies in the evaluation of how the presence of strategic interaction between future improved technology generations will impact on the adoption timing of a present, intermediate technology.

Our study proceeds as follows. In Section 2 we present our general stochastic duopoly model of sequential technologies with the feature of strategic interaction between future and improved technologies. In Section 3 we characterize the upgrading of the intermediate technology. Section 4 characterizes the timing of adoption of the intermediate technology with a particular emphasis on how spillovers to the rival will impact on this timing. Finally, we offer some concluding comments in Section 5.

2   A Model of Strategic Adoption of an Intermediate Technology

Consider the strategic interaction between a firm with access to an intermediate technology, the leader (L), and a rival firm, the follower (F), with a potential to enter the market with a future improved version of the technology. Initially, the intermediate technology (I) generates a flow of revenues $\pi_L^{I,0} : \mathsf{R}_+ \mapsto \mathsf{R}$. This revenue flow accruing to L depends on a stochastic state variable, which evolves over time according to a process to be specified below. Further, the experience accumulated through the adoption of the intermediate technology opens the option of updating the intermediate technology (I) to a superior and final technology (U). The timing of arrival of the updated technology is stochastic. More precisely, the updated technology becomes available to L with an arrival time τ_L which is exponentially distributed with a parameter λ_L. Thus, adoption of the intermediate technology opens up an imbedded real option of a stochastic technology improvement governed by a Poisson process with the parameter λ_L. At the moment of updating its intermediate technology the irreversible cost for L is given by c_L. The experience accumulated through L's adoption of the intermediate technology will also generate spillover effects. Thereby F benefits stochastically in the sense that L's adoption of the intermediate technology opens up the opportunities for F to acquire the final technology. The final technology becomes available to F with an arrival time τ_F. This arrival time is exponentially distributed with a parameter λ_F satisfying the property that $\lambda_F \leq \lambda_L$. Consequently, L's adoption the intermediate technology initiates a Poisson process for both L and F with respect to the arrival date of the updated technology version and the intensity of this Poisson process is higher for L than for F so as to capture the natural feature that L is able to benefit from the intermediate technology to a higher extent than F. However, the stochastic nature of the arrival processes implies that it could perfectly well happen that F succeeds to adopt the updated technology earlier than L.

3

Especially, we observe that

$$\mathsf{P}[\tau_L \leq \tau_F] = \frac{\lambda_L}{\lambda_L + \lambda_F} \geq \frac{\lambda_F}{\lambda_L + \lambda_F} = \mathsf{P}[\tau_L \geq \tau_F] \quad \text{and} \quad E[\tau_L] \quad \frac{1}{\lambda_L} \leq \frac{1}{\lambda_F} = E[\tau_F].$$

Consequently, we find that the closer is λ_F to λ_L, the closer the expected arrival dates are to each other. Analogously, the closer is λ_F to λ_L, the closer is the probability $1/2$ of the event that F adopts the updated technology earlier than L. In the limit where neither of the parties dominates stochastically the updating opportunity and $\lambda_F = \lambda_L$, we have that $\mathsf{P}[\tau_L \leq \tau_F] = \mathsf{P}[\tau_L \geq \tau_F] = 1/2$ and $E[\tau_F] = E[\tau_L]$. The joint probability distributions determining the expected order of adoption are presented in detail in section 3. At the moment of adopting the final technology the irreversible cost for F is given by c_F. Again we assume L's experience with the intermediate technology to translate into an adoption cost advantage so that $c_L \leq c_F$. At each point in time the revenue flow is determined by the prevailing market structure. Firstly, we let $\pi^{i,j}_L$ denote the revenue flow accruing to L when the state variable is x and when (i, j) summarizes the state of the technology such that $i \in \{I, U\}$ indicates whether L makes use of the intermediate (I) or the updated technology (U), whereas $j \in \{0, U\}$ describes whether F has entered with the updated technology (U) or whether it is still inactive (0). Secondly, we define $\pi^{i,j}_F(x)$ as the revenue flow of F when the state variable is x and when now (i, j) summarizes the state of the technology such that $i \in \{0, U\}$ indicates whether F has entered with the updated technology (U) or whether it is still inactive (0), whereas $j \in \{0, U\}$ delineates the technology adopted by L. In particular, $\pi^{I}_L(x)$ denotes the flow of monopoly revenues to L associated with the intermediate technology, whereas analogously $\pi^{U,0}_L(x)$ captures the monopoly revenues for the updated technology. Given these assumptions, it is clear that the cash flow of the leader now reads as

$$\pi_L(t, x, \bar{\tau}) = \begin{cases} \pi^{I,0}_L(x)1_{[0,\tau_L)}(t) + \pi^{U,0}_L(x)1_{(\tau_L,\tau_F]}(t) + \pi^{U,U}_L(x)1_{(\tau_F,\infty)}(t), & \tau_L \leq \tau_F \\ \pi^{I,0}_L(x)1_{[0,\tau_F)}(t) + \pi^{I,U}_L(x)1_{(\tau_F,\tau_L]}(t) + \pi^{U,U}_L(x)1_{(\tau_L,\infty)}(t), & \tau_L \geq \tau_F, \end{cases} \quad (2.1)$$

where $\bar{\tau} = (\tau_L, \tau_F)$, and

$$1_A(x) = \begin{cases} 1, & x \in A \\ 0, & x \notin A \end{cases}$$

denotes the indicator (or characteristic) function of the set A. We make the following assumption regarding the relationship between the market structure and the revenue flows.

Assumption 2.1. *For all $x \in \mathsf{R}_+$ the following relationships hold:*

$$\pi^{U,0}_L(x) > \pi^{I,0}_L(x) > \pi^{I,U}_L \quad (2.2)$$

$$\pi^{U,0}_L(x) > \pi^{U,U}_L(x) > \pi^{I,U}_L(x) \quad (2.3)$$

$$\pi^{U,I}_F(x) > \pi^{U,U}_F(x) \quad (2.4)$$

$$\pi^{U,U}_L(x) = \pi^{U,U}_F(x). \quad (2.5)$$

Assumption (2.2) states the natural idea that the updated technology is superior to the intermediate technology in the sense that it yields higher monopoly revenues. Furthermore, successful adoption on behalf of the follower is assumed to burden the profits of the leader. Assumptions (2.3) and (2.4) formalize the notion that successful implementation of the updated technology yields competitive advantages. Assumption (2.5) means that the two rivals are symmetric in the product market in case they have both succeeded to implement the updated technology. In Assumption 2.1 the relationships (2.2) and (2.3) imply that there are first mover and second mover advantages, respectively. In line with, for example, Reinganum (1981) and Stenbacka and Tombak (1994) we assume the first mover advantages to dominate relative to the second mover advantages. This is formally expressed by

Assumption 2.2. *Define the mapping* $\Delta : \mathsf{R}_+ \mapsto \mathsf{R}$ *by*

$$\Delta(x) = \pi_L^{U,0}(x) - \pi_L^{I,0}(x) + (\pi_L^{U,U}(x) - \pi_L^{I,U}(x)) \tag{2.6}$$

This mapping satisfies the inequality $\Delta(x) > 0$ *for all* $x \in (0, \infty)$.

Finally, throughout our study we require that the following assumption holds.

Assumption 2.3. *The profit flows* $\pi_L^{i,j}(x)$ *and* $\pi_F^{i,j}(x)$ *are monotonically increasing as functions of the current state* x.

This assumption implies, in particular, that the cash flow of a firm is increasing independently of the order at which the upgrading dates τ and τ are realized. It should, however, be emphasized that the order of the upgrading dates, τ_F and τ, will in an essential way affect the speed whereby the profit flows increase as functions of x. F

We assume that the underlying state variable evolves according to a linear, time homogeneous, and regular diffusion process defined on the complete filtered probability space $(\Omega, P, \{\mathcal{F}_t\}_{t\geq0}, \mathcal{F})$ and evolving on the state-space R_+ according to the dynamics described by the Itô-stochastic differential equation

$$X(0) = x, \tag{2.7}$$

where both the drift coefficient $\mu : \mathsf{R}_+ \mapsto \mathsf{R}$ and the diffusion coefficient $\sigma : \mathsf{R}_+ \mapsto \mathsf{R}$ are assumed to be sufficiently smooth for guaranteeing the existence and uniqueness of a solution for the stochastic differential equation (2.7) (at least continuous, cf. Borodin and Salminen (1996), pp. 46–47), and $W(t)$ denotes standard Brownian motion. We assume that this process is independent of the adoption timings. In order to avoid interior singularities, we also assume that the diffusion coefficient $\sigma(x)$ is positive, that is, we assume that $\sigma(x) > 0$ for all $x \in (0, \infty)$. Moreover, since we are not interested in potential default we further assume that the boundaries 0 and ∞ of the state-space are natural for the process $X(t)$. Given the diffusion $X(t)$, denote now as $\mathcal{L}^1(\mathsf{R}_+)$ the class of measurable mappings satisfying the absolute integrability condition (i.e the *absence of speculative bubbles condition*)

$$E_x \int_0^\infty e^{-rt} |f(X(t))| dt < \infty.$$

For any $f \in \mathcal{L}^1(\mathsf{R}_+)$ we define the functional $R_r f : \mathsf{R}_+ \mapsto \mathsf{R}$ as

$$(R_r f)(x) = E_x \int_0^\infty e^{-rt} f(X(t)) dt.$$

Moreover, it is well-known that given the assumptions of our study, there are two linearly independent *fundamental solutions* $\psi(x)$ and $\varphi(x)$ satisfying a set of appropriate boundary conditions based on the boundary behavior of the process X and spanning the set of solutions of the ordinary differential equation $((\mathcal{A} - r)u)(x) = 0$ (cf. Borodin and Salminen, pp. 18 - 19). Moreover,

$$\frac{\psi'(x)}{S'(x)}\varphi(x) - \frac{\varphi'(x)}{S'(x)}\psi(x) = B,$$

where $B > 0$ denotes the constant Wronskian of the fundamental solutions $\psi(x)$ and $\varphi(x)$ and

$$S'(x) = \exp\left(-\int \frac{2\mu(x)dx}{\sigma^2(x)}\right)$$

denotes the density of the scale function of X.

5

3   Upgrading of the Intermediate Technology

In order to derive an explicit representation of the real option value of the leader's opportunity to update its technology at τ_L we have to derive the probability distributions characterizing the joint probability distributions of the dates τ_L and τ_F given that the order of the outcomes are known. These distributions and their densities are derived in Appendix A of this paper. Define now the random variable $D(\tau_F, \tau_L) = \tau_F - \tau_L$ measuring the difference between the arrival dates of the upgraded technology. It is clear that given our observations in the previous section, the expected difference

$$E[D(\tau_F, \tau_L)] = \frac{\lambda_L - \lambda_F}{\lambda_F \lambda_L} > 0$$

is increasing as a function of the intensity λ_L and decreasing as a function of λ_F. Moreover, given the joint probability distributions of the dates τ_L and τ_F given that the order of the outcomes are known, we find that

$$E[D(\tau_F, \tau_L); \tau_L \leq \tau_F] = \frac{\lambda_L}{\lambda_F(\lambda_F + \lambda_L)}$$

and

$$E[D(\tau_F, \tau_L); \tau_L > \tau_F] = -\frac{\lambda_F}{\lambda_L(\lambda_F + \lambda_L)}.$$

Consequently, we find that the expected duration of future monopoly benefits accruing to the leader is an increasing function of the intensity λ_L and a decreasing function of λ_F. Similarly, we observe that the expected duration of the benefits for the leader to catch up given that the follower has upgraded his technology first is a decreasing function of λ_L and an increasing function of λ_F.

Given the joint probability distributions of the arrival dates τ_L and τ_F when the order of the outcomes are known, we find that the real option value can be expressed as

$$V_L(x) = E_x \int_0^\infty e^{-rs} \pi_L(s, X(s), \bar{\tau}) ds = \Gamma_{L,1} + \Gamma_{L,2}, \tag{3.1}$$

where elementary calculus can be applied to establish that

$$\Gamma_{L,1} = \frac{\lambda_L}{\lambda_L + \lambda_F}(R_r \pi_L^{I,0})(x) + E_x\left[e^{-r\tau_L}(R_r(\pi_L^{U,0} - \pi_L^{I,0}))(X_{\tau_L}); \tau_L < \tau_F\right]$$
$$+ E_x\left[e^{-r\tau_F}(R_r(\pi_L^{U,U} - \pi_L^{U,0}))(X_{\tau_F}); \tau_L < \tau_F\right]$$

and

$$\Gamma_{L,2} = \frac{\lambda_F}{\lambda_L + \lambda_F}(R_r \pi_L^{I,0})(x) + E_x\left[e^{-r\tau_F}(R_r(\pi_L^{I,U} - \pi_L^{I,0}))(X_{\tau_F}); \tau_L > \tau_F\right]$$
$$+ E_x\left[e^{-r\tau_L}(R_r(\pi_L^{U,U} - \pi_L^{I,U}))(X_{\tau_L}); \tau_L > \tau_F\right].$$

In (3.1) the term $\Gamma_{L,1}$ measures the expected cumulative present value of the cash flow accrued by the leader given that the leader receives the upgraded technology prior to the follower and $\Gamma_{L,2}$ measures the expected cumulative present value of the cash flow accrued by the leader when the opposite happens. Given these findings, we have been able to demonstrate that the real option value of the leader reads as

$$V_L(x) = (\lambda_L + \lambda_F)(R_{r+\lambda_L+\lambda_F}(R_r\Delta))(x) + \lambda_F(R_{r+\lambda_F}(R_r(\pi_L^{U,U} - \pi_L^{U,0})))(x)$$
$$+ (R_r\pi_L^{I,0})(x) + \lambda_L(R_{r+\lambda_L}(R_r(\pi_L^{U,U} - \pi_L^{I,U})))(x), \tag{3.2}$$

where $\Delta(x) > 0$, as we have assumed the first mover advantages to dominate relative to the second mover advantages (Assumption 2.2). It is worth pointing out that the assumed monotonicity of

6

the cash flows imply that that the expected cumulative present value of the cash flow $\pi_L(t,x,\bar{\tau})$ is increasing as well (since expectation preserves ordering) and, therefore, that the real option value $V_L(x)$ of the leader is increasing as a mapping of the current state x. It is also worth emphasizing that according to the classical resolvent equation, we have under the assumptions of our study that for any $\lambda \in \mathsf{R}_+$ and $f \in \mathcal{L}^1(\mathsf{R}_+)$

$$(R_r f)(x) - (R_{r+\lambda} f)(x) = \lambda(R_{r+\lambda}(R_r f))(x).$$

Consequently, we find that the real option value $V_L(x)$ can be rewritten as

$$V_L(x) = (R_r \pi_L^{U,U})(x) - (R_{r+\lambda_L+\lambda_F}\Delta)(x) - (R_{r+\lambda_F}(\pi^{U,U}_L - \pi^{U,0}_L \qquad (3.3)$$

From (3.2) we can conclude that the real option value of the leader reads as

$$V_L(x) = (R_r \pi_L^{I,0})(x) + \lambda_L(R_{r+\lambda_L}(R_r(\pi_L^{U,0} - \pi_L^{I,0})))(x) \qquad (3.4)$$

whenever $\lambda_F \downarrow 0$, that is, whenever the follower can never upgrade its technology. As is clear from (3.3), the real option value (3.4) can also be expressed as

$$V_L(x) = (R_r \pi_L^{U,0})(x) - (R_{r+\lambda_L}(\pi^{U,0}_L - \pi^{I,0}_L))(x)$$

This real option value captures the sum of the current production potential (in terms of the current incumbent technology) and the expected cumulative present value accrued from upgrading the incumbent intermediate technology, i.e., the expected cumulative present value of the first mover advantage (which is sure when $\lambda_F \downarrow 0$ since then the follower is never expected to upgrade).

Another important extreme case is the case where the leader is expected to be able to adopt instantaneously the upgraded technology, that is, when $\lambda_L \uparrow \infty$. In that case the real option value of the leader reads as

$$V_L(x) = (R_r \pi_L^{U,0})(x) + \lambda^F(R_{r+\lambda^F}(R_r(\pi_L^{U,U} - \pi_L^{U,0})))(x) \qquad (3.5)$$

The reason for this finding is obvious. If $\lambda_L \uparrow \infty$, then the leader expects that the upgraded technology can be adopted immediately and, consequently, that the follower can never be first. Again, (3.3) implies that the real option value (3.5) can in this case also be rewritten as

$$V_L(x) = (R_r \pi_L^{U,U})(x) - (R_{r+\lambda_F}(\pi^{U,U}_L - \pi^{U,0}_L))(x).$$

In order to characterize the comparative static properties of the real option value $V_L(x)$ more precisely, we first present the next auxiliary result:

Lemma 3.1. *The real option value $V_L(x)$ is continuously differentiable with respect to the intensities λ_L and λ_F. Moreover, for all $x \in \mathsf{R}_+$ we have*

$$\frac{\partial V_L}{\partial \lambda_L}(x) = -\frac{\partial}{\partial \lambda_L}(R_{r+\lambda_L+\lambda_F}\Delta)(x) - \frac{\partial}{\partial \lambda_L}(R_{r+\lambda_L}(\pi_L^{U,U} - \pi_L^{I,U}))(x) > 0,$$

and

$$\frac{\partial V_L}{\partial \lambda_F}(x) = -\frac{\partial}{\partial \lambda_F}(R_{r+\lambda_L+\lambda_F}\Delta)(x) - \frac{\partial}{\partial \lambda_F}(R_{r+\lambda_F}(\pi_L^{U,U} - \pi_L^{U,0}))(x) < 0.$$

Proof. See Appendix B. $\square$

Lemma 3.1 establishes that an increase in the intensity λ_L will unambiguously increase the value of the real option value $V_L(x)$. Similarly, Lemma 3.1 also establishes that an increase in the intensity λ_F unambiguously decreases the option value $V_L(x)$.

7

4 Adoption of the Intermediate Technology

Adoption of the intermediate technology represents a real option available to the leader. With k_L denoting the irreversible cost of adopting the intermediate technology the value of this real option is given by

$$J_L(x) = \sup_\tau E_x \left[e^{-r\tau}(V_L(X(\tau)) - k_L) \right], \tag{4.1}$$

which incorporates the upgrading opportunity embedded in $V_L(x)$. Before presenting a set of necessary conditions under which the valuation problem (4.1) can be solved explicitly, we first state a set of generally valid conclusions characterizing the comparative statics of the value of the optimal adoption policy. In order to accomplish this task, we denote as $\hat{J}_L(x)$ and $\hat{V}_L(x)$ the values in the presence of the intensity $\hat{\lambda}_L$ satisfying the inequality $\hat{\lambda}_L \geq \lambda_L$. Similarly, we denote as $\tilde{J}_L(x)$ and $\tilde{V}_L(x)$ the values in the presence of the intensity $\tilde{\lambda}_F$ satisfying the inequality $\tilde{\lambda}_F \geq \lambda_F$. Our main results characterizing the comparative statics[1] of the value of the optimal adoption policy are now presented in

Theorem 4.1. *An increase in the intensity λ_L (λ_F) increases (decreases) the value $J_L(x)$. More precisely, $\hat{J}_L(x) \geq J_L(x)$ and $\tilde{J}_L(x) \leq J_L(x)$ for all $x \in \mathsf{R}_+$.*

Proof. See Appendix C. $\qquad\qquad\qquad\qquad\qquad\qquad\qquad\qquad\qquad\qquad\qquad\qquad\qquad$ $\square$

Theorem 4.1 shows that the sign of the relationship between the intensity λ_L and the value $J_L(x)$ is unambiguously positive under the conditions underlying this theorem. Thus, an improvement in the quality of the leader's updating option will increase the value of the embedded option associated with the intermediate technology. In a similar manner, by speeding up the arrival of the updated technology to the follower an increase in the upgrading intensity λ_F impacts negatively on the option value of the intermediate technology to the leader, since the increased spillovers to the follower reduce the probability of monopoly profits for the leader.

In light of the well-known properties of the Poisson distribution it is worth pointing out that the conclusions above can equally well be interpreted in terms of the uncertainty affecting the intensities of updating. More precisely, an increase in the intensity λ_L decreases not only the expected arrival date of the updated technology, but it simultaneously decreases the variance λ_L^{-2} and, thus, the uncertainty of the arrival of the updated technology. Consequently, an increase in the uncertainty of the firm-specific updating process will increase the option value of the intermediate technology. On the other hand, we can conclude that an increase in the uncertainty of the updating opportunities facing the rival will decrease the option value of the leader's intermediate technology. In fact, we can interpret this as a negative strategic impact of the uncertainty incorporated in the follower's updating opportunity on the option value of the leader's intermediate technology.

A set of generally satisfied sufficient conditions under which the optimal adoption problem is explicitly solvable is now summarized in the following.

Theorem 4.2. *Assume that the mapping $H(x) = (V_L(x) - k_L)/\psi(x)$ attains a unique global maximum at $x^* = \operatorname{argmax}\{H(x)\} \in \mathsf{R}_+$, and that $(AV_L)(x) \leq r(V_L(x) - k_L)$ for all $x \in (x^*, \infty)$. Then, the optimal adoption date is $\tau(x^*) = \inf\{t \geq 0 : X(t) \geq x^*\}$ and the value of adopting the intermediate technology is*

$$J_L(x) = \psi(x) \sup_{y \geq x}[H(y)] = \begin{cases} V_L(x) - k_L, & x \geq x^* \\ H(x^*)\psi(x), & x < x^*. \end{cases} \tag{4.2}$$

---

[1]It should be emphasized that the real option value of adopting the intermediate technology will satisfy ordinary comparative static properties with respect to the mean and variance of the underlying diffusion X.

8

Especially,

$$V_L'(x^*)\psi(x^*) = (V_L(x^*) - k_L)\psi'(x^*)$$

and the value $J_L \in C^1(\mathsf{R}_+) \cap C^2(\mathsf{R}_+ \backslash \{x^*\})$ *satisfies for* $x \in \mathsf{R}_+$ *the variational inequalities*

$$\min\{((r - \mathcal{A})J_L)(x), J_L(x) - (V_L(x) - k_L)\} = 0. \tag{4.3}$$

Proof. See Appendix D. □

Theorem 4.2 defines a set of sufficient conditions under which the optimal adoption problem has a unique solution. In order to develop an intuitive explanation of Theorem 4.2, assume that the leader decides to adopt the intermediate technology whenever the diffusion $X(t)$ exceeds a given arbitrary state $y \in \mathsf{R}_+$ implying that the adoption date is $\tau(y) = \inf\{t \geq 0 : X(t) \geq y\}$. The resulting value of such policy is

$$G(x, y) = \begin{cases} {}_L(x) - k_L & x \geq y \\ (V_L(y) - k_L)\frac{\psi(x)}{\psi(y)} & x < y \end{cases}$$

The potential sub-optimality of this class of decision rules clearly implies that $G(x, y) \leq J_L(x)$. Theorem 4.2 essentially specifies a set of conditions under which the value of adopting the intermediate technology can be expressed in terms of the auxiliary mapping $G(x, y)$ as

$$J_L(x) = G(x, x^*).$$

Thus, the optimal value of adopting the intermediate technology can be attained by choosing that state at which $G(x, y)$ is maximized as the adoption threshold. It is worth emphasizing that the ordinary first order condition $G_y(x, x^*) = 0$ then implies the smooth-pasting condition $V_L'(x^*)\psi(x^*) = (V_L(x^*) - k_L)\psi'(x^*)$.

In order to characterize the sign of the relationship between the updating intensities and the threshold x^* at which the intermediate technology is optimally adopted, we now define the mapping

$$H(x, \lambda_L, \lambda_F) = \frac{V_L(x)}{\psi(x)}.$$

We can now establish the following:

Corollary 4.3. *Assume that the conditions of Theorem 4.2 are satisfied.*

(i) If $H_{x\lambda_L}(x, \lambda_L, \lambda_F) < 0$ *for all* $x < y$ *and* $\lambda_L > \lambda_F$, *then* $\frac{\partial x^*}{\partial \lambda_L} < 0$.

(ii) If $H_{x\lambda_F}(x, \lambda_L, \lambda_F) > 0$ *for all* $x < y$ *and* $\lambda_L > \lambda_F$, *then* $\frac{\partial x^*}{\partial \lambda_F} > 0$.

Proof. See Appendix E. □

Corollary 4.3 presents a set of conditions under which we can unambiguously determine the sign of the relationship between the updating intensities and the threshold x^* at which the intermediate technology is optimally adopted. Especially, we find that the sign of this relationship is essentially determined by the sensitivity of the mapping $\Delta(x) = V_L'(x)\psi(x) - V_L(x)\psi'(x)$ with respect to changes in the updating intensities. It is worth emphasizing that although the effect of a change in one of the updating intensities to the real option value $V_L(x)$ is known (from Lemma 3.1), the effect on the marginal value $V_L'(x)$ is not known. The result of Corollary 4.3 dealing with the impact of an increase in λ_L is illustrated in the following figure.

9

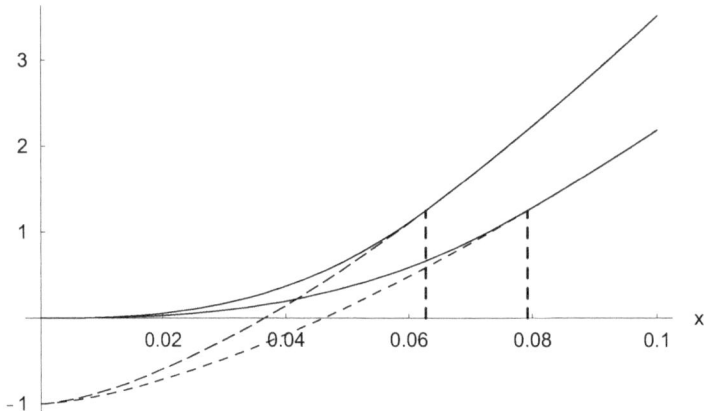

Figure 1: The impact of an increase on λ_L to $V_L(x)$ and $J_L(x)$

4.1 Explicit Processes: An Illustration

In order to illustrate the general results incorporated in Theorem 4.2 we now focus on an environment where the stochastic process as well as the cash flows are made explicit. More precisely, we assume the conditions (A) and (B) to hold.

(A) The controlled process evolves according to a geometric Brownian motion. This means that $X(t)$ constitutes the solution of the stochastic differential equation

$$dX(t) = \mu X(t)dt + \sigma X(t)dW(t) \quad X(0) = x, \tag{4.4}$$

where both the drift coefficient $\mu \geq 0$ and the volatility coefficient $\sigma > 0$ are exogenously given constants.

(B) The cash flows are of the form $\pi_L^{i,j}(x) = a_L^{i,j} x^\theta$, where $a_L^{i,j} \in \mathsf{R}_+$ and $\theta \in \mathsf{R}_+$ are known constants.

It is well-known that under the circumstances defined by (A) and (B), we have that

$$(R_r \pi_L^{i,j})(x) = \frac{a_L^{i,j} x^\theta}{r - \delta(\theta)}$$

provided that the absence of speculative bubbles condition (i.e. absolute integrability condition) $r > \delta(\theta) = \mu\theta + \sigma^2\theta(\theta-1)/2$ is met. Similarly, we observe that

$$(\lambda_L + \lambda_F)(R_{r+\lambda_L+\lambda_F}(R_r\Delta))(x) = (\lambda_L + \lambda_F)\frac{(a_L^{U,0} - a_L^{I,0} - a_L^{U,U} + a_L^{I,U})x^\theta}{(r - \delta(\theta))(r + \lambda_F + \lambda_L - \delta(\theta))}, \tag{4.5}$$

$$\lambda_F(R_{r+\lambda_F}(R_r(\pi_L^{U,U} - \pi_L^{U,0})))(x) = \lambda_F\frac{(a_L^{U,U} - a_L^{U,0})x^\theta}{(r - \delta(\theta))(r + \lambda_F - \delta(\theta))}, \tag{4.6}$$

and

$$\lambda_L(R_{r+\lambda_L}(R_r(\pi_L^{U,U} - \pi_L^{I,U})))(x) = \lambda_L\frac{(a_L^{U,U} - a_L^{I,U})x^\theta}{(r - \delta(\theta))(r + \lambda_L - \delta(\theta))}. \tag{4.7}$$

10

Consequently, we find that the real option value reads as

$$V_L(x) = \frac{A(\theta, \lambda_L, \lambda_F)x^\theta}{r - \delta(\theta)}, \qquad (4.8)$$

where

$$A(\theta, \lambda_L, \lambda_F) = a_L^{I,0} + (\lambda_L + \lambda_F)\frac{(a_L^{U,0} - a_L^{I,0} - a_L^{U,U} + a_L^{I,U})}{(r + \lambda_F + \lambda_L - \delta(\theta))}$$
$$+ \lambda_F\frac{(a_L^{U,U} - a_L^{U,0})}{(r + \lambda_F - \delta(\theta))} + \lambda_L\frac{(a_L^{U,U} - a_L^{I,U})}{(r + \lambda_L - \delta(\theta))}.$$

Standard differentiation of the multiplier $A(\theta, \lambda_L, \lambda_F)$ with respect to the intensity λ_L now yields that

$$\frac{\partial A}{\partial \lambda_L}(\theta, \lambda_L, \lambda_F) = (r - \delta(\theta))\left[\frac{(a_L^{U,0} - a_L^{I,0} - a_L^{U,U} + a_L^{I,U})}{(r + \lambda_F + \lambda_L - \delta(\theta))^2} + \frac{(a_L^{U,U} - a_L^{I,U})}{(r + \lambda_L - \delta(\theta))^2}\right] > 0, \qquad (4.9)$$

since $a_L^{U,0} - a_L^{I,0} > a_L^{U,U} - a_L^{I,U}$ and $a_L^{U,U} > a_L^{I,U}$ by assumption. That is, increasing the intensity at which the updated technology can be adopted increases the real option value $V_L(x)$. Differentiating the multiplier $A(\theta, \lambda_L, \lambda_F)$ now with respect to the intensity λ_F yields

$$\frac{\partial A}{\partial \lambda_F}(\theta, \lambda_L, \lambda_F) = (r - \delta(\theta))\left[\frac{(a_L^{U,0} - a_L^{I,0} - a_L^{U,U} + a_L^{I,U})}{(r + \lambda_F + \lambda_L - \delta(\theta))^2} + \frac{(a_L^{U,U} - a_L^{U,0})}{(r + \lambda_F - \delta(\theta))^2}\right] < 0, \qquad (4.10)$$

since it was assumed that $a_L^{I,U} < a_L^{I,0}$ (assumption (2.2)). In other words, increasing the intensity λ_F decreases the real option value $V_L(x)$. It is now an elementary exercise to establish that $A(\theta)$ can be rewritten as

$$A(\theta, \lambda_L, \lambda_F) = \frac{(r - \delta(\theta))}{(r + \lambda_L + \lambda_F - \delta(\theta))}a_L^{I,0} + \frac{(r - \delta(\theta))\lambda_L}{(r + \lambda_L + \lambda_F - \delta(\theta))(r + \lambda_F - \delta(\theta))}a_L^{U,0}$$
$$+ \frac{(r - \delta(\theta))\lambda_F}{(r + \lambda_L + \lambda_F - \delta(\theta))(r + \lambda_L - \delta(\theta))}a_L^{I,U}$$
$$+ \frac{\lambda_L\lambda_F(2(r - \delta(\theta)) + \lambda_L + \lambda_F)}{(r + \lambda_L + \lambda_F - \delta(\theta))(r + \lambda_L - \delta(\theta))(r + \lambda_F - \delta(\theta))}a_L^{U,U}$$

which is positive by assumption. The value of the real investment opportunity to adopt the intermediate technology is now given by

$$V_L(x) - k_L = \frac{A(\theta, \lambda_L, \lambda_F)x^\theta}{r - \delta(\theta)} - k_L.$$

Consequently, we have been able to establish the following.

Theorem 4.4. *Assume that $r > \delta(\theta)$. Then, the optimal investment date is $\tau^* = \inf\{t \geq 0 : X(t) \geq x^*(\lambda_L, \lambda_F)\}$, where*

$$x^*(\lambda_L, \lambda_F) = \left(1 - \frac{\theta}{\eta_2}\right)^{1/\theta}\left(\frac{rk_L}{A(\theta, \lambda_L, \lambda_F)}\right)^{1/\theta}$$

denotes the optimal investment threshold and

$$\eta_2 = \frac{1}{2} - \frac{\mu}{\sigma^2} - \sqrt{\left(\frac{1}{2} - \frac{\mu}{\sigma^2}\right)^2 + \frac{2r}{\sigma^2}} < 0$$

11

denotes the negative root of the characteristic equation $\sigma^2 b(b-1)/2 + \mu b - r = 0$ of the ordinary (Cauchy-Euler) differential equation $\sigma^2 x^2 u''(x)/2 + \mu x u'(x) - ru(x) = 0$. The value of the optimal policy reads as

$$
\begin{aligned}
J_L(x) &= x^{\eta_1} \sup_{y \geq x} \left[y^{-\eta_1} \left(\frac{A(\theta, \lambda_L, \lambda_F) y^\theta}{r - \delta(\theta)} - k_L \right) \right] \\
&= \begin{cases} V_L(x) - k_L & x \geq x^*(\lambda_L, \lambda_F) \\ x^{\eta_1} x^*(\lambda_L, \lambda_F)^{-\eta_1} (V_L(x^*(\lambda_L, \lambda_F)) - k_L) & x < x^*(\lambda_L, \lambda_F) \end{cases},
\end{aligned}
$$

where

$$
\eta_1 = \frac{1}{2} - \frac{\mu}{\sigma^2} + \sqrt{\left(\frac{1}{2} - \frac{\mu}{\sigma^2} \right)^2 + \frac{2r}{\sigma^2}} > 0
$$

denotes the positive root of the characteristic equation stated above.

Proof. The alleged result follows from Theorem 4.2. $\qquad\square$

From Theorem 4.4 we can immediately infer that $\partial x^*(\lambda_L, \lambda_F)/\partial \lambda_L < 0$ from which we can conclude that an increase in the intensity λ_L will decrease the optimal investment threshold and thereby speed up the adoption of the intermediate technology. Analogously, we find that $\partial x^*(\lambda_L, \lambda_F)/\partial \lambda_F > 0$, which means that an improvement in the follower's ability to exploit the upgraded technology will raise the leader's threshold for investing into the intermediate technology. Thus, an improvement in the follower's ability to transform the intermediate technology into an improved technology will slow down the leader's adoption of the intermediate technology. We summarize these properties in

Corollary 4.5. *The threshold of adopting the intermediate technology depends negatively (positively) on the leader's (follower's) upgrading intensity.*

In light of Theorem 4.2 and Corollary 4.5 we can conclude that the incentives of adopting an intermediate technology is highly related to the upgrading intensities of the leader and the follower, respectively. In fact, we can make the interpretation of the relationship $\nu = \lambda_F/\lambda_L$ as a measure of the technological spillover from L to F created by L's investment into the intermediate technology. In particular, $\nu = 0$ captures the case of complete intellectual property protection such that L would be able to maintain exclusivity with respect to applications of the innovation incorporated in the intermediate technology. At the other end of the spectrum $\nu = 1$ serves as a representation of complete spillovers as exemplified by open-source access to particular software technologies.

In light of Corollary 4.5 it is interesting to investigate iso-incentive curves defined by $x^*(\lambda_L, \lambda_F) = K$, where $K > 0$ is an arbitrary constant. Such iso-incentive curves describe the set of updating intensities (λ_L, λ_F) which keep the adoption threshold constant. By substituting the definition of the iso-incentive curve into the definition of the optimal adoption threshold $x^*(\lambda_L, \lambda_F)$ we find that

$$
A(\theta, \lambda_L, \lambda_F) = \frac{r k_L (\eta_2 - \theta)}{\eta_2 K^\theta}
$$

has to hold along the iso-incentive curve. Consequently, we find that the slope of the iso-incentive curve $x^*(\lambda_L, \lambda_F) = K$ is

$$
\frac{\partial \lambda_L}{\partial \lambda_F}\bigg|_{x^*(\lambda_L, \lambda_F) = K} = - \frac{A_{\lambda_F}(\theta, \lambda_L, \lambda_F)}{A_{\lambda_L}(\theta, \lambda_L, \lambda_F)} > 0
$$

because of (4.9) and (4.10). In fact, by comparing (4.9) and (4.10) we can conclude that the relative magnitudes of the partial derivatives $A_{\lambda_F}(\theta, \lambda_L, \lambda_F)$ and $A_{\lambda_L}(\theta, \lambda_L, \lambda_F)$ are essentially related to those of the *catch-up benefit*, $a_L^{U,U} - a_L^{I,U}$, and the *loss to be caught-up*, $a_L^{U,0} - a_L^{U,U}$.

12

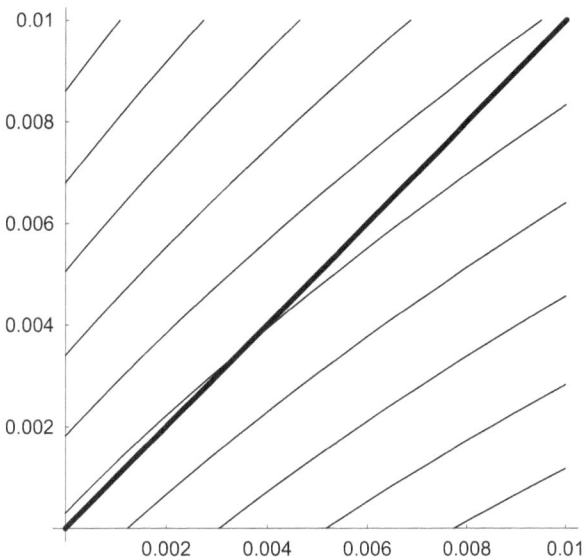

Figure 2: Iso-incentive curves when $\theta = 1.5$

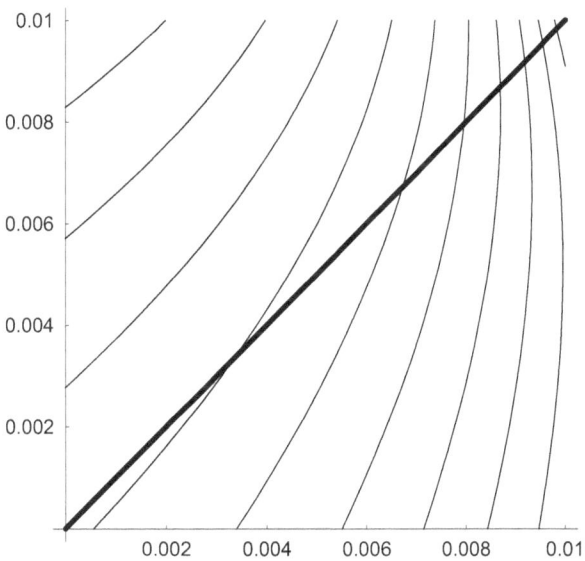

Figure 3: Iso-incentive curves when $\theta = 3.5$

The form of the iso-incentive curves are illustrated in Figure 1 for moderately convex and in Figure 2 for strongly convex cash flows.

In his recent review of high-technology industries, Varian (2001) views versioning as a characteristic feature of these innovation intense industries. According to Varian business strategies

13

based on versioning offer a mechanism for intertemporal discrimination. Versioning can also be viewed in light of the present model. Our analysis can be seen to have its focus on the following question. How are the incentives of adopting an intermediate technology related to the strategic relationship prevailing between the updating capabilities of competing firms in the product market? A high degree of exclusivity (in the sense of a low ν-ratio), which could be achieved, for example, through an auctioning mechanism like in the UMTS-auctions, implies adoption of a fairly premature intermediate technology characterized by a low upgrading intensity λ_L. Thus, by inducing early adoption of intermediate technologies a high degree of exclusivity would promote versioning in the sense of prolonging the expected phase of intermediate technologies.

In light of this reasoning our model analytically captures the relationship between the degree of exclusivity incorporated in future technology generations and the timing of adoption of intermediate technology versions. Such a relationship seems to have clear implications for policies determining the intellectual property protection, like patent policy. Thus, our model could be seen to offer an analytical framework for how to evaluate patent policies in an intertemporal context which emphasizes the dynamics of technological progress through successive generations of improved technologies. In the present context we leave it for future studies to explore these policy implications in greater detail.

5 Concluding comments

In this study we have applied a real options approach to analytically characterize the option value of adopting an intermediate technology. We formally designed an asymmetric duopoly model to delineate how the optimal adoption timing of an intermediate technology depends on the embedded upgrading options available to the firm itself (the leader) and to its future rival (the follower). Focusing on diffusions we developed explicit representations demonstrating that the threshold of adopting an intermediate technology depends negatively (positively) on the leader's (follower's) upgrading intensity.

We showed that an improvement in the quality of the leader's updating option will increase the value of the embedded option associated with the intermediate technology, whereas an increase in the follower's upgrading intensity was demonstrated to impact negatively. Furthermore, an increase in the uncertainty of the firm-specific updating process was proved to increase the option value of the intermediate technology, whereas the uncertainty incorporated in the follower's updating opportunity was shown to have a negative strategic impact on the option value of the leader's intermediate technology.

For the case of an underlying stochastic process following the geometric Brownian motion we explicitly characterized how competition taking place with respect to future technology generations will impact on the timing of adopting an intermediate technology. This was carried out in our analysis of the iso-incentive curves keeping the leader's incentives of adopting the intermediate technology invariant.

The present model can be used as a benchmark for important generalizations in several dimensions. Firstly, throughout the analysis the upgrading intensities have been assumed to be exogenous features of the model. Of course, in a more complete model of dynamic technology competition these updating intensities should result from optimizing investment decisions at each point in time. Such an extension would be particularly important for extensions focusing on implications for technology policy or intellectual property protection. The restriction to exogenous upgrading intensities might represent a much smaller loss of generality in contexts, like ours, emphasizing merely first approximations of the dynamics of technological progress through successive generations of improved technologies.

Secondly, our model has admittedly represented an artificial short-cut as an analysis of strategic interaction insofar as it was restricted to a one-sided perspective. In this respect a complete analysis should focus on timing equilibria of games of strategic adoption of intermediate technologies where symmetric duopolists would identical opportunities to adopt an intermediate

14

technology. Such an approach would generate a theory of strategic compound options. In a world where we still do not have an well-established theory of strategic options such an approach would represent a highly ambitious research task.

Acknowledgements: The financial support from the *Foundation for the Promotion of the Actuarial Profession (Aktuaaritoiminnan Kehittämissäätiö)* to Luis H. R. Alvarez is gratefully acknowledged. Both authors acknowledge the financial support from the *Yrjö Jahnsson Foundation*.

References

[1] Alvarez, L. H. R. *Reward functionals, salvage values, and optimal stopping*, 2001, *Mathematical Methods of Operations Research*, **54**, 315–337.

[2] Alvarez, L. H. R. and Stenbacka, R. *Adoption of Uncertain Multi-stage Technology Projects: A Real Options Approach*, 2001, *Journal of Mathematical Economics*, **35**, 71–97.

[3] Borodin, A. and Salminen, P. *Handbook on Brownian motion - Facts and formulae*, 1996, Birkhauser, Basel.

[4] Dixit, A. K. and Pindyck, R. S. *Investment under uncertainty*, 1994, Princeton UP, Princeton.

[5] Øksendal, B. *Stochastic differential equations: An introduction with applications*, (Fourth Edition) 1995, Springer, Berlin.

[6] Perotti, E. and Rossetto, S. *Strategic Advantage and the Optimal Exercise of Entry Options*, 2001, *Centre for Economic Policy Research (CEPR)*, Discussion Paper No. 3061.

[7] Reinganum, J., *On the Diffusion of New Technology: A Game Theoretic Approach*, 1981, *Review of Economic Studies*, **48**, 395–405.

[8] Stenbacka, R. and M. Tombak, *Strategic Timing of Adoption of New Technologies under Uncertainty*, 1994, *International Journal of Industrial Organization*, **12**, 387–411.

[9] Trigeorgis, L. *Real Options: Managerial flexibility and strategy in resource allocation*, 1996, MIT Press.

[10] Varian, H., *High-Technology Industries and Market Structure*, 2001, In **Economic Policy for the Information Economy**, Proceedings of a Symposium Sponsored by the Federal Reserve Bank of Kansas City, Jackson Hole, Wyoming August 30 - September 1, 2001.

15

A   The joint probability densities

The assumed independence of the arrival dates of the upgraded technology implies that the joint (unconditional) distribution of the arrival dates is of the form $\lambda_L \lambda_F\, e^{-\lambda_L \tau_L - \lambda_F \tau_F}$. Consequently, ordinary integration yields that the joint probability distributions of the dates τ_L and τ_F given that the order of the outcomes are known are:

$$P_L(\alpha; \tau_L < \tau_F) = \mathsf{P}[\tau_L \leq \alpha; \tau_L < \tau_F] = \frac{\lambda_L}{\lambda_L + \lambda_F}\left(1 - e^{-(\lambda_L + \lambda_F)\alpha}\right),$$

$$P_{L-}(\alpha; \tau_L > \tau_F) = \mathsf{P}[\tau_L \leq \alpha; \tau_L > \tau_F] = 1 - e^{-\lambda_L \alpha} - \frac{\lambda_L}{\lambda_L + \lambda_F}\left(1 - e^{-(\lambda_L + \lambda_F)\alpha}\right),$$

$$Q(\alpha; \tau_L > \tau_F) = \mathsf{P}[\tau_F \leq \alpha; \tau_L > \tau_F] - \frac{\lambda_F}{\lambda_L + \lambda_F}\left(1 - e^{-(\lambda_L + \lambda_F)\alpha}\right),$$

and

$$Q_{-}(\alpha; \tau_L < \tau_F) = \mathsf{P}[\tau_F \leq \alpha; \tau_L < \tau_F] = 1 - e^{-\lambda_F \alpha} - \frac{\lambda_F}{\lambda_L + \lambda_F}\left(1 - e^{-(\lambda_L + \lambda_F)\alpha}\right).$$

The continuous differentiability of these distributions implies that the densities of these probability distributions can be expressed as

$$p_L(\alpha; \tau_L < \tau_F) = \frac{d}{d\alpha}\mathsf{P}[\tau_L \leq \alpha; \tau_L < \tau_F] = \lambda_L e^{-(\lambda_L + \lambda_F)}$$

$$p_L(\alpha; \tau_L > \tau_F) = \frac{d}{d\alpha}\mathsf{P}[\tau_L \leq \alpha; \tau_L > \tau_F] = \lambda_L e \left(1 - e^{-\lambda_F \alpha}\right),$$

$$q_L(\alpha; \tau_L > \tau_F) = \frac{d}{d\alpha}\mathsf{P}[\tau_F \leq \alpha; \tau_L > \tau_F] = \lambda_F e^{-(\lambda_L + \lambda_F}$$

and

$$q_L(\alpha; \tau_L < \tau_F) = \frac{d}{d\alpha}\mathsf{P}[\tau_F \leq \alpha; \tau_L < \tau_F] = \lambda_F e^{-\lambda_F}\left(1 - e^{-\lambda_L \alpha}\right).$$

Given these densities, it now involves nothing but ordinary integration to establish that for any $\in \mathcal{L}^1(\mathbf{R}_+)$, that is, for any cash flow $f(x)$ having a finite expected cumulative present value, we have that

$$E_x\left[e^{-r\tau_L}f(X_{\tau_L}); \tau_L < \tau_F\right] = \lambda_L(R_{r+\lambda_L+\lambda_F}f)(x),$$

$$E_x\left[e^{-r\tau_L}f(X_{\tau_L}); \tau_L > \tau_F\right] = \lambda_L(R_{r+\lambda_L}f)(x) - \lambda_L(R_{r+\lambda_L+\lambda_F}f)(x),$$

$$E_x\left[e^{-r\tau_F}f(X_{\tau_F}); \tau_L < \tau_F\right] = \lambda_F(R_{r+\lambda_F}f)(x) - \lambda_F(R_{r+\lambda_F+\lambda_L}f)(x),$$

and

$$E_x\left[e^{-r\tau_F}f(X_{\tau_F}); \tau_L > \tau_F\right] = \lambda_F(R_{r+\lambda_L+\lambda_F}f)(x).$$

B   Proof of Lemma 3.1

Proof. The continuous differentiability of the real option value $V_L(x)$ with respect to the intensities λ_L and λ_F follows directly from (3.3) and the continuous differentiability of the exponential function. The positivity of $\partial V_L/\partial \lambda_F$ follows directly from the positivity of the mappings $\Delta(x)$ and $\pi_L^{U,U}(x)$, $\pi^{I,U}(x)$ and the negativity of the impact of increased intensity on the discount factors $e^{-(r+\lambda_L)s}L$ and $e^{-(r+\lambda_L+\lambda_F)s}$. In order to prove the negativity of $\partial V_L/\partial \lambda_L$ we first observe that $\Delta(x) = \pi_L^{U,0}(x) - \pi^{I,0}(x) - (\pi_L^{U,U}(x) - \pi_L^{I,U}(x)) - \pi_L^{U,0}(x) - \pi_L^{U,U}(x)$ for all $x \in \mathbf{R}_+$. Consequently, we find that

$$e^{-(r+\lambda_L+\lambda_F)s}s\Delta(X(s)) \leq e^{-(r+\lambda_L)s}s\Delta(X(s)) \leq e^{-(r+\lambda_L)s}s(\pi_L^{U,0} - \pi_L^{U,U})(X(s))$$

completing the proof of our lemma. $\qquad\qquad\square$

16

C   Proof of Theorem 4.1

Proof. As is shown in [5] (p. 200) the increasing sequence

$$J_n(x) = \sup_{t \geq 0} E_x \left[e^{-rt} J_{n-1}(X(t)) \right], \quad J_0(x) = V_L(x) - k_L$$

converges to the value function $J_L(x)$. Similarly, the increasing sequence

$$\hat{J}_n(x) = \sup_{t \geq 0} E_x \left[e^{-rt} \hat{J}_{n-1}(X(t)) \right], \quad \hat{J}_0(x) = \hat{V}_L(x) - k_L$$

converges to the value function $\hat{J}_L(x)$. As was shown in Lemma 3.1 we have that $\hat{V}_L(x) - k_L \geq V_L(x) - k_L$ implying that $\hat{J}_L(x) \geq \hat{J}_n(x) \geq J_n(x)$ from which the inequality $\hat{J}_L(x) \geq J_L(x)$ follows by monotone convergence. The proof of the second claim is analogous. □

D   Proof of Theorem 4.2

Proof. Denote as $J_L^*(x)$ the value of the proposed stopping strategy. It is now clear that since the proposed stopping time τ^* is admissible and

$$J_L^*(x) = E_x \left[e^{-r\tau(x^*)} (V_L(X(\tau(x^*))) - k_L) \right]$$

we have that $J_L^*(x) \leq J_L(x)$. To prove the opposite inequality, we first observe that $J_L^* \in C^1(\mathbf{R}_+) \cap C^2(\mathbf{R}_+ \backslash \{x^*\})$, that

$$\lim_{x \uparrow x^*} J_L^{*''}(x) = H(x^*) \psi''(x^*) < \infty \quad \lim_{x \downarrow x^*} J_L^{*''}(x) = V_L''(x^*) < \infty,$$

that $J_L^*(x) \geq (V_L(x) - k_L)^+$ for all $x \in \mathbf{R}_+$, and that $(\mathcal{A}J_L^*)(x) \leq rJ_L^*$ for all $x \in \mathbf{R}_+ \backslash \{x^*\}$. Thus, $J_L^*(x)$ constitutes a r-excessive majorant of the exercise payoff $V_L(x) - k_L$. However, since $J_L(x)$ is the least of these majorants, we find that $J_L^*(x) \geq J_L(x)$, thus proving that $J_L^*(x) = J_L(x)$. The validity of the variational inequality (4.3) follows directly from the definition (4.2) of the value function. □

E   Proof of Corollary 4.3

Proof. Since

$$\frac{\partial H}{\partial x}(x, \lambda_L, \lambda_F) = \frac{V_L'(x)\psi(x) - V_L(x)\psi'(x)}{\psi^2(x)},$$

we find that

$$\frac{\partial^2 H}{\partial x \partial \lambda_L}(x, \lambda_L, \lambda_F) = \frac{\frac{\partial V_L'}{\partial \lambda_L}(x)\psi(x) - \frac{\partial V_L}{\partial \lambda_L}(x)\psi'(x)}{\psi^2(x)}$$

and

$$\frac{\partial^2 H}{\partial x \partial \lambda_F}(x, \lambda_L, \lambda_F) = \frac{\frac{\partial V_L'}{\partial \lambda_F}(x)\psi(x) - \frac{\partial V_L}{\partial \lambda_F}(x)\psi'(x)}{\psi^2(x)}.$$

On the other hand, implicit differentiation of the optimality condition $V_L'(x^*)\psi(x^*) = (V_L(x^*) - k_L)\psi'(x^*)$ yields that

$$[V_L''(x^*)\psi(x^*) - (V_L(x^*) - k_L)\psi''(x^*)]\frac{\partial x^*}{\partial \lambda_L} = \psi'(x^*)\frac{\partial V_L}{\partial \lambda_L}(x^*) - \psi(x^*)\frac{\partial V_L'}{\partial \lambda_L}(x^*)$$

17

and that

$$[V_L''(x^*)\psi(x^*) - (V_L(x^*) - k_L)\psi''(x^*)]\frac{\partial x^*}{\partial \lambda_F} = \psi'(x^*)\frac{\partial V_L}{\partial \lambda_F}(x^*) - \psi(x^*)\frac{\partial V_L'}{\partial \lambda_F}(x^*).$$

The optimality of the threshold x^* implies that $V_L''(x^*)\psi(x^*) < (V_L(x^*) - k_L)\psi''(x^*)$. Combining this expression with our assumptions completes the proof of the alleged results. $\qquad\square$

18

Analysis Template for Assessing e-Business Investments by Applying Real Options

DR. ANETT MEHLER-BICHER

EUROPEAN BUSINESS SCHOOL (ebs)
Schloss Reichartshausen
D 65375 Oestrich-Winkel

e mail: anett.bicher@ebs.de

Abstract

The application of option theory to problems in information technology and especially to e-business has been the subject of some research in the last years. The theoretical advantage of option pricing models – namely to derive different options and assess them by basing well-founded mathematical models – at the same time reflects its practical disadvantage. As a consequence, the goal of this research paper is to provide a useful best practice. It focuses on systemizing the process of deriving and structuring feasible options as well as on facilitating the election of the option model and the estimation of the input variables by providing an analysis template. Structure and content of this template will be introduced in this paper as well as a demonstration of its practical feasibility will be given.

Motivation

The application of option theory to problems in information technology and especially to e-business has been the subject of some research in the last years. Prior research (Benaroch 2000) especially targeted the demonstration of the power of real options through applying fundamental option pricing models, such as the Black-Scholes or the binomial models, on real world cases. As a result, option theory offers high potential for useful insights regarding the evaluation of irreversible investments under uncertainty and requiring flexibility.

The theoretical advantage of option pricing models – namely to derive different options and assess them by basing well-founded mathematical models – at the same time reflects its practical disadvantage. Especially mathematical complexity and lack of practical experience are major problems. In order to establish option pricing models in practice, the user needs adequate tools not

1

only for calculating concrete numbers, but also for structuring the investment problem into feasible options and for deriving the corresponding input variables.

As a consequence, the goal of this research paper is to provide a useful best practice. It focuses on systemizing the process of deriving and structuring feasible options as well as on facilitating the election of the option model and the estimation of the input variables. An essential means is an analysis template, which significantly supports this process through specific questioning and serves as a management instrument for applying option pricing in practice. For demonstrating its feasibility the analysis template is applied to a real world case.

Structure and content of this template will be introduced in this paper. The structure is oriented towards a three-tier evaluation process, namely identification, assessment and active management of the option portfolio. Therefore, the template mainly consists of three modules each corresponding to a tier in the evaluation process; module 1 and 2 will be introduced in the scope of this paper. The first module serves as a systematical derivation of the option; existing option alternatives will be visualized graphically for making the discussion of further option alternatives feasible. By defining essential premises the establishment of the optimal option pricing model is enabled. High vividness resulting in less calculation exactness, which does not exhaust the flexibility completely, or an exact reflection of the reality in the model are questions to be clarified. After specifying the model the actual parameter estimation starts; this is rendered by a systematical approach given in the second module.

This research paper makes three important contributions in this context: (1) it systemizes the application of real option models by providing a best practice, (2) it allows the application of option theory by giving practitioners an analysis template at hand and (3) it presents the first application of the analysis template that uses a real world case.

General Aspects of the Analysis Template

Real Option Valuation Cycle

Ideally a real option pricing (ROP) method is an integrative element of a three-tier management cycle oriented towards the shareholder value concept. The three tiers comprehend "identification", "assessment" and "active management of the option portfolios" (see fig. 1).

Figure 1

Fig. 1: RO valuation cycle.

Identification, i.e. structuring the problem adequately and determining the different option alternatives of the investment is a very demanding task and very often more problematic than assessing the different options. Thus, this research work focuses on both the identification and the assessment part. The goal is to systemize the process of deriving and structuring feasible options as well as on facilitating the election of the option model and the estimation of the input parameters. An analysis template supporting the management to identify the option alternatives as well as to estimate the input variables systematically will be an essential means for the ROP's practicability. Due to complexity reasons for enabling a first application of option pricing the analysis template focuses on "identification" as well as "assessment"; future research work will concentrate on extending the analysis template regarding "active management of the option portfolios".

Goal and Structure of the Analysis Template

The analysis template, which is oriented to a specific need, i. e. to find out and obtain the relevant information/data to process a real option valuation, is designed as a questionnaire. On behalf of the complexity of ROP it is intended not to let the respondents work alone with the analysis template, but to apply the template in guided interviews or workshops in order to ideally support an RO valuation.

A high-qualitative questionnaire (Saldern 1998; Kirchhoff 2001; Konrad 2001) has to (a) be oriented towards the respondent and his/her needs, (b) be oriented towards the goal of valuation, (c) consist of clear instructions on how to proceed, and (d) ease the evaluation of answers. Additionally, (a) questions to unknown problems/facts have to be explained by a pre-example, (b) the context has to be explained, (c) every answer alternative has to be understandable or the option of not answering has to be provided, (d) questions have to be oriented to the knowledge and experience of the respondent and (e) the questionnaire has to consist of acceptable questions. Besides the content the layout is very important; different topic sections have to be indicated. Furthermore, the questionnaire has to start with an introduction consisting of instructions and the goal of the valuation examination.

Providing an analysis template in order to ideally support an RO valuation means that the template has to be tested to find out where the pitfalls are and if the questionnaire is applicable. However, although follow-up research will be necessary to test the questionnaire intensively, in the scope of this paper a pre-test is proceeded by applying the template to the case of Yahoo PayDirect. The analysis template has to be seen as a guide for the difficult valuation process; the outcome will be a valuation of the investment project's managerial flexibilities, and therefore, a realistic, non-static value of the investment project. Furthermore, the respondent will gain a better understanding of and a better estimation about the value of the managerial flexibilities. The complete questionnaire is divided into the three modules "identification", "assessment" and "active management" each consisting of multiple sub-modules. The respondent can process one module, and then he/she will receive a certain outcome from it. Afterwards, he/she can decide if he/she wants to continue with the next module.

Basic instruction elements are used to systemize the questionnaire and to facilitate its usage. Usually, instructions precede each section. Whereas the text field with two types of grey color ⬛ indicates necessary information, the text field with the bright middle color effects ⬛ indicates information that the respondent has to fill out. A line indicates a write-in value or information.

Module 1: Identification

Justification of ROP Application

On behalf of the characteristics of e-business projects and their corresponding investments, verifying whether the application of ROP is justified is more than less a formal check. E-business projects and related investments are always characterized by high uncertainty, irreversibility and great demand of flexibility. Nevertheless, analyzing the kind and structure of the planned project helps to classify the project's subject with regard to risk and uncertainty. An appropriate classification of e-business projects requires a systematic scheme to place the planned application. A common classification is the categorization between companies based on an existing business model and those developing a new business model (Amram 2000). Refining this approach by identifying the main uncertainty drivers (Mehler-Bicher 2002)

- "product/service",
- "business model",
- "market",

- "technology",

- "enterprise/brand"

as well as distinguishing between new or existing ones allows to develop a profile for categorizing e-business applications (see fig. 2).

In the analysis template justifying the ROP application takes place by briefly describing the project and specifying the project profile (see fig. 2). By analyzing these five dimensions and specifying the planned project profile by connecting the five extensions, the degree of uncertainty and risk can be determined qualitatively. One principle conclusion can be drawn: The bigger the hatched area in the profile, the more the increase in risk and uncertainty and the more difficult fluctuations are to predict.

Figure 2

Fig. 2: Justification of ROP application.

Option Derivation

Assessing and managing options firstly requires their identification and derivation. Although it is rational not to analyze all options possible of a certain investment, a statement about the probable value of an option cannot be determined in all cases in advance. In contrast to financial options, another difficulty is that a precise specification of real options, which are usually clustered and mutually effecting, often cannot take place. Thus, an adequate structuring of the problem often is of greater importance than an exact and detailed calculation.

In practice, first off the base case scenario DCF model usually has to be stated. Secondly, based on this model feasible options with a significant effect on the project value have to be derived. Despite the variety of thinkable options, different option types can be classified (Trigeorgis 1998; Amram 1999; Hommel 1999a; Copeland 2000). The knowledge of the limited number of option types having a significant effect on the project value is an essential means for identifying and defining relevant options. In concrete situations the option types will be designed or instantiated differently. In most cases compound options representing the inter- or intra-project interaction occur and require the explicit consideration of the correlation among the different option parts.

By analyzing the initial situation in order to specify the base case scenario DCF model (see fig. 3) and to derive feasible options a concentration on the two or three most probable options is a promising strategy. On one hand the analysis effort is limited, on the other hand follow-up calculations will be characterized by high informative value. As a consequence, for practical purposes supporting the derivation of the two to three most feasible as well as probable options is intended by the analysis template.

Figure 3

Fig. 3: Derivation of base case scenario.

Due to specific option elements depending on the option type, a separate questionnaire sheet has been developed for each option. After having identified the option type a five step process has to be initiated for each option. Firstly the option has to be described. Secondly a strategic analysis of the option takes place; exclusiveness of ownership, project interaction and urgency of decision have to be specified. Thirdly, the relation between the option and the base case scenario has to be visualized by developing an option tree (see fig. 4 exemplary for Option to Wait).

Figure 4

Fig. 4: Option Derivation: Option to Wait (part 1).

Fourthly, the option needs have to be characterized. The degree of involved uncertainty and market loss potential as well as the potential of increase in information and risk reduction by applying a specific option are essential aspects. After having determined the lifetime of the option the overall importance of the specific option has to be graded qualitatively. Finally, specific option type elements have to be specified (see fig. 5 exemplary for Option to Wait).

Figure 5

Fig. 5: Option Derivation: Option to Wait (part 2).

Model Election

Not only a large number of investment approaches, but also a huge variety of option pricing models, which can be roughly distinguished in analytical and numerical procedures, exists. In e-business the analytical model deployed by SCHWARTZ and ZOZAYA-GOROSTIZA (Schwartz 2000) and the numerical capability-based approach by KULATILAKA (KULATILAKA 1996) are the two most elaborated ROP methods. Although specific models exist, due to complexity their applicability is not unproblematic.

The results established by the application of option pricing and their appropriateness as well as exactness mainly depend on the specific approach. High vividness resulting in less calculation exactness, which does not exhaust the flexibility completely, or an exact reality's reflection of the model are questions to be clarified. A comparison of the three mostly used models – binomial model, log-transformed binomial model and Black-Scholes – demonstrates the advantages of the binomial model. Besides its clear and simple structure as well as its vividness the binomial model is convincing due to its high variability and flexibility related to the structure of the input parameters. An application is feasible for discrete as well as continuous input parameters or simple as well as compound options.

In certain situations, i. e. situations where a statement about the structure of the input variables is difficult to establish, the strategy to specify the option model after having estimated the parameters is advantageous. Due to the paper's scope the model election part is restricted to the binomial model, the log-transformed binomial procedure and the BLACK-SCHOLES formula (see fig. 6). The comparison given in MEHLER-BICHER (Mehler-Bicher 2001) allows the determination of a feasible option model; future research work will focus on extending this part by considering further option models.

Figure 6

Fig. 6: Model election.

Module 2: Assessment

Parameter Estimation

The input variables required for assessing an option are (a) underlying (V), (b) exercise price (X), (c) time to maturity (t), (d) risk-free interest rate (r_f), (e) dividends (δ) and (f) volatility (s). Besides the construction of the underlying the estimation of the volatility is the most elaborative task, especially because slightly changing the volatility leads to significant differences in the resulting option pricing calculation.

For estimating the input variables, first off the analysis template focuses on deriving the values of underlying, exercise price, time to maturity and risk-free interest rate (see fig. 7).

Figure 7

Fig. 7: Parameter estimation (part 1).

Then the questionnaire concentrates on the estimation of dividends and volatility (see fig. 8).

Figure 8

Fig. 8: Parameter estimation (part 2).

Calculation

Using specialized software packages simplifies and reduces the calculation effort; especially a graphical visualization of different options supports the managerial decision process. Since supporting this task by an analysis template requires a means for selecting an appropriate software tool and an adaptation to fast-changing circumstances, this module is not presented within this paper.

Exemplary Application of the Analysis Template

Case Introduction

Yahoo, Inc., is a global Internet company offering a comprehensive branded network of services worldwide. Brought to life in 1994, Yahoo became one of the first online navigational guides to the World Wide Web. Today, with nearly 90 million unique visitors each month Yahoo is a leading web portal in terms of traffic, advertising, and household reach. Starting out as a search engine and link directory Yahoo has since added a host of additional features to its web services. Advancing its "Yahoo Everywhere!" strategy, the company has extended its reach by providing content and services for wireless and handheld devices. In February, 2000, in a move intended to sharpen its e-commerce profile, Yahoo acquired Arthas.com, a provider of online person-to-person (P2P) payment services. Despite the obvious advantages of online P2P payment systems, previous virtual currency services have failed to gain acceptance among consumers. In November 1999 PayPal launched its service for Palm devices and email and was quickly followed by a number of competitors. When Yahoo acquired Arthas.com, dotBank was one of three online P2P payment systems, that had been already been launched. At that time, 16,000 individuals were already using PayPal, dotBank posted about 8,000 users.

Figure 9

Fig. 9: Concept of Yahoo PayDirect (Source: Yahoo Inc.)

After the acquisition, Yahoo decided to take dotBank offline until the service could be properly integrated into the Yahoo family of services. The system was scheduled for re-launch under its new name Yahoo PayDirect (see fig. 9) in May 2000. However, the integration process took until July 31, when PayDirect was officially launched. In the meantime, PayPal had already signed up to 2.7 million users and was adding new customers at a rate of 20,000 per day. By December 2000,

8

PayPal had surpassed the 5 million user mark and was clearly evolving as the winning P2P payment service in the US. By September 2000, it had become obvious to Yahoo management that they would not be able to stop PayPal on its way to dominate P2P solutions in the US. As a consequence, issues in the international roll-out of the service had to be discussed. Since timing was critical and not all countries were equally attractive for P2P payment solutions, Yahoo decided to launch PayDirect in the UK.

Case Evaluation

An exemplary a-posteriori application of the analysis template to the case of Yahoo PayDirect serves for demonstrating its feasibility. On behalf of the restricted scope of this paper and the sensitivity of the data, only selected parts focusing on module 1 will be presented.

Justification of ROP Application

The three constituting elements of option pricing models – uncertainty, irreversibility and flexibility – are certainly fulfilled. A lot of parameters influencing the project's value cannot be precisely projected; in addition it is not possible for Yahoo to recoup any of its costs for business and technical development effort or legal expenses. Yahoo's management requires a very high degree of flexibility including e. g. the option to limit losses by simply abandoning the product or by stopping advertising activities.

Analyzing Yahoo PayDirect and specifying its profile – new service, new business model, new market, existing technology, existing enterprise/brand – leads to the result that Yahoo PayDirect is a highly risky and uncertain application (see fig. 10).

Figure 10

Fig. 10: Specifying the profile of Yahoo PayDirect.

Option Derivation

The first step to derive feasible options is to state the base case scenario without any optionality (see fig. 11).

Figure 11

Fig. 11: Base case scenario for Yahoo PayDirect.

Yahoo possessed a rather high degree of management flexibility in their decision to launch PayDirect in the UK. In order to analyze the size of the opportunity and to come to a conclusion about the decision to launch, they discussed feasible alternatives. Since the competitors would take at least 6 months to concur the UK market, postponing the launch would allow Yahoo to get a better feeling about the market, its potential and the competitors. Should they decide to launch right now and things go bad, Yahoo still has the option to abandon the product altogether at any point in time, defaulting on planned future marketing spending. In contrast, if Yahoo is able to establish a sizable user base in UK quickly enough, highly profitable extensions of the original product could be introduced in the future. The most important follow-on service would involve cross-border foreign exchange (FX) payments or another interesting growth option would involve wireless devices and mobile commerce.

A detailed analysis of Yahoo PayDirect shows that three significant options exist. Alongside a wait option (Option A see fig. 12)

Figure 12

Fig. 12: Option A for Yahoo PayDirect.

two alternatives have to be considered:

- Option B:

  After launch re-evaluate the project every half year and freeze marketing spending if the project doesn't take off.

- Option C:

  After launch wait two years and then start possible service expansions only if the economics support their launch.

Combining the three options leads to the following option tree (see fig. 13).

Figure 13

Fig. 13: Option tree for Yahoo PayDirect.

Model Election

Focusing on option A allows to apply all three option pricing models (see fig. 14). However, option B and C require the application of the binomial model or a log-transformed binomial procedure. For simplification reasons it is assumed that volatility, exercise price and risk-free interest rate are constant and determinable. In addition, the assumptions are made that dividends happen in discrete payments and price evolution process follows a geometric Brownian process.

Figure 14

Fig. 14: Model election for Yahoo PayDirect.

Conclusion and Outlook

Due to complexity the practical application of ROP procedures requires appropriate tools. The analysis template presented serving as a best practice focuses on supporting the identification part as well as the assessment part and thus represents a valuable means for identifying most relevant aspects for correctly applying option pricing models. In addition, the mostly very time-consuming process of ROP application can be significantly shortened. Especially the exemplary application of the analysis template has shown significant support.

On top of gaining more detailed questionnaire modules future research work will focus on the extension of the model election; the restriction to selected models has to be skipped and the module has to be generalized. Another research aspect will be considered in module 3: active management of the option portfolio. For gaining additional significant insights applying this analysis template to multiple case studies will be the next major task.

References

Amram, M., Kulatilaka, N. (1999). "Disciplined Decisions: Aligning strategy with the Financial Mar-ketes." Harvard Business Review 77(1): 95-104.

Amram, M., Kulatilaka, N. (2000). "Strategy and Shareholder Value Creation." Journal of Applied Corpo-rate Finance 13(2): 8-21.

Benaroch, M. K., R. (2000). "Justifying Electronic Banking Network Expansion Using Real Options Analysis." MIS Quarterly 24(2): 197-225.

Copeland, T., Koller, T., Murrin, J. (2000). Valuation: Measuring and Managing the Value of Companies. New York, Wiley.

Hommel, U., Pritsch, G. (1999a). Investitionsbewertung und Unternehmensführung mit dem Realoptionsansatz. Handbuch Corporate Finance: Konzepte, Strategien, und Praxiswissen für das moderne Finanzmanagement. A. Achleitner, Thoma, G.F. Köln. 1-68.

Kirchhoff, S., Kuhnt, S., Lipp, P., Schlawin, S. (2001). Fragebogen - Datenbasis, Konstruktion und Auswertung. Opladen, Leske + Budrich.

Konrad, K. (2001). Mündliche und schriftliche Befragung - Ein Lehrbuch. Landau, Verlag Empirische Pädagogik.

Kulatilaka, N., Balasubramanian, P., Storck, J. (1996). Managing Information Technology Investments: A Capability-based Real Options Approach. Boston, MA, Working Paper 96-35, Boston University.

Mehler-Bicher, A. (2001). Questionnaire Template for Assessing e-Business Investments by Applying Option Pricing. Eighth European Conference on IT Evaluation, Oxford.

Mehler-Bicher, A. (2002). Evaluating e-Business Investments By Defining Profiles. submitted to: Nineth European Conference on IT Evaluation, Paris.

Saldern, M. (1998). Befragung und Beobachtung im Betrieb. Baltmannsweiler, Schneider Verlag Hohengehren.

Schwartz, E., Zozaya-Gorostiza, C. (2000). Valuation of Information Technology Investments as Real Options, Working Paper, Anderson Graduate School of Management.

Trigeorgis, L. (1998). Real Options: Managerial Flexibility and Strategy in Ressource Allocation. Cambridge, MA.

DAY 3

SESSION 2

Matts Rosenberg
"Capital Investment and Labor Adjustment Under
Uncertainty – Empirical Evidence from Finland"

*Christer Carlsson
"Industry Foresight"

Svante Olofsson
"Agent Applications"

*No Material

Capital Expenditures and Labor Demand under Uncertainty

Matts Rosenberg[*]

Graduate School of Finance and Financial Accounting

Swedish School of Economics and Business Administration

Helsinki, Finland

This Draft

March 8, 2002

(Draft version, please do not cite or circulate without permission.)

Keywords: Capital expenditures, Labor demand, Uncertainty

JEL classification: D21, D81, G31, J23

[*] Contact information: Swedish School of Economics and Business Administration, P.O. Box 479, 00101 Helsinki, Finland. E-mail: rosenber@shh.fi, Tel. +358 (0)9 4313 3473, Fax +358 (0)9 4313 3393. The author is indebted to Eva Liljeblom, Anders Löflund, Henrik Palmén, Daniel Pasternack, Rune Stenbacka, and the participants at the GSFFA Joint Finance Research Seminar for valuable comments and discussions. Financial support from Stiftelsen Svenska Handelshögskolan is gratefully acknowledged.

Capital Expenditures and Labor Demand under Uncertainty

Abstract

This paper analyzes the effect of uncertainty on capital expenditures and labor demand based on Finnish panel data during the time period 1987 – 2000. The paper contributes to existing empirical research regarding firm behavior under uncertainty, by extending the firm's investment problem to a dual-factor framework, including both capital and labor. Utilizing a stock return based measure of uncertainty decomposed into systematic and idiosyncratic components, and employing a Generalized Method of Moments (GMM) estimation procedure, the results reveal that idiosyncratic uncertainty significantly reduces capital expenditures. However, it seems as if this effect is driven by the latter part of the investigation period (1994 – 2000). The results fail to find any significant relationship between capital expenditures and the effects of total and systematic uncertainty. In the case of labor demand, I find a modest negative effect of total uncertainty on labor demand. Systematic and idiosyncratic uncertainty does not affect the desired employment level of the firm.

I. Introduction

Firms operating in dynamic economic environments are exposed to uncertainty regarding the evolution of the underlying state variables. The relationship between investment and uncertainty is far from resolved in the economic literature. Leahy and Whited (1996) classify theories of investment under uncertainty along two dimensions. The first dimension is manifested by models that examine the firm in isolation, emphasizing the uncertainty of some distinct state variable in the firm's environment, and secondly, of models analyzing the firm in a market context. In the case of isolation, uncertainty, *per se*, matters for investment behavior. When the firm is analyzed in a market context, uncertainty matters only through the channel of covariance in the returns between investment projects.[1] Along the other dimension we find models that analyze investment behavior contingent on the functional form of the marginal revenue product of capital. This category consists of models predicting that the marginal revenue product of capital is convex in some random variable, and models that predict that the relationship is concave. In the former case, increased uncertainty would be associated with increased investment activity, whereas a concave form would imply a depressing effect on investment activity.

In the models of Hartman (1972) and Abel (1983), the marginal revenue product of capital is a convex function of shocks facing the firm. Thus, increased uncertainty is associated with an increase in the marginal unit of capital, which increases the firm's incentive to invest.[2] Models predicting a concave relationship of the marginal revenue product of capital are generally referred to as models of irreversible investment (McDonald and Siegel (1986), Pindyck (1988), Dixit and Pindyck (1994). The irreversible investment framework relies on the assumption that firms hold investment (call) options on the expected returns generated by investment projects. Thus, returns to investment are asymmetric, due to the fact that the firm has an option to wait for uncertainty to be resolved before the option to invest is exercised, or not. The implication of these models is that increased uncertainty should be associated with

---

[1] The effect of covariance is manifested, e.g., in the capital asset pricing model (CAPM). The effect on increased uncertainty (risk), measured as the covariance of the returns of an investment project with respect to market returns, would according to the CAPM-framework reduce the incentive to invest, due to an increase in the required rate of return on the investment.

[2] In general, a convex form of the marginal revenue product of capital is a result of Jensen's inequality, which implies that a mean preserving spread in the distribution of a shock facing the firm will increase the expected return on investment.

decreased investment activity. Later work by Abel, Dixit, Eberly, and Pindyck (1996), have generalized the basic model of irreversible investment by including the effect of divestment (put) options. In the generalized case, the comparative static effect of uncertainty on investment becomes ambiguous, and depends on the interactions of the firm's call and put options.[3]

Economic theory defines the firm's production as a function of capital and labor, $Y = f(K, L)$. The above-presented models, however, focus on investment in capital and make the assumption that labor is flexible. In the models of Hartman (1972) and Abel (1983) it is exactly the flexibility of labor relative to capital that produces a convex shape of the marginal revenue product of capital. Characterizing capital as quasi fixed, in the sense that it must be chosen before uncertainty is resolved, and treating labor as flexible may not be an ideal assumption. Many issues affect the employment decisions of the firm in the short-term, such as adjustment costs related to hiring and firing, firm specific investment in labor, labor unions and labor intensity, which are related to the elasticity of labor costs (Oi 1962). Hartman (1976) empasizes the importance of the assumptions regarding the nature of flexibility of factor inputs. For instance, assuming that the firm is forced to choose both of its inputs (capital and labor) before uncertainty is resolved will in fact reduce the demand for both factors. Hence, depending on the chosen assumptions a wide range of theoretically supportable outcomes are possible, and it is, *a priori*, difficult to make predictions about the relationship between uncertainty and the demand for the firm's input factors. Thus, the current paper contributes to existing empirical research regarding firm behavior under uncertainty, by extending the firm's investment problem to a dual-factor framework, including both capital and labor.

Empirical evidence regarding capital expenditures has been documented by Leahy and Whited (1996), who examined the investment – uncertainty relationship for a panel of U.S. firms. Utilizing a measure of uncertainty based on stock returns, the results revealed that an increase in uncertainty decreases firm investment, primarily through its effect on Tobin's Q. Furthermore, the results revealed that a CAPM-based risk measure

---

[3] However, a number of recent empirical surveys have argued in favor of the basic model of irreversible investment, and hence, emphasizing the role of the firm's call options compared to the associated put options, due to factors such as capital specificity and "lemons problems" (Akerlof 1970).

did not show any significance with respect to firm investment. Bo and Sterken (2000) conducted a survey based on a sample of Dutch firms. The employed methodology involved specification of the threshold value of profits that theoretically should trigger firm investment. The results revealed that firms on average were concerned with the option values of investment opportunities in making investment decisions, i.e., they tend to wait until actual profits reaches its threshold level. Henley, Carruth, and Dickerson (2000) conducted a survey for a panel of UK firms, where uncertainty was specified both as a sectoral risk measure and as an idiosyncratic risk measure. Sectoral uncertainty was found to significantly depress firm investment, whereas idiosyncratic uncertainty was found to increase investment. In addition, the results revealed that the effect of uncertainty was larger in absolute magnitude for diversified firms compared to focused firms. Bloom, Bond, and Van Reenen (2001) examined the investment behavior of a sample of UK firms. Following Leahy and Whited (1996), uncertainty was measured using the standard deviation of stock returns. The main prediction was supported by the data, as the results revealed that the response of investment to demand increases was lower when uncertainty was high. Finally, Bulan (2001) examined the relationship between firm investment and uncertainty for a panel of U.S. firms. In line with previous empirical work, uncertainty was measured as the volatility of the firm's stock returns. In addition, total uncertainty was decomposed into systematic, industry, and idiosyncratic components. The results revealed that uncertainty significantly reduces firm investment. Furthermore, the survey included an examination of the hypotheses that market competition erodes the option value of waiting, and that size affects the investment – uncertainty relationship.[4] The results supported both hypotheses by revealing that more competitive firms were less responsive to industry uncertainty compared to less competitive firms, and that large firms exhibited greater sensitivity to uncertainty compared to small firms.

Labor policy surveys have specified the determinants of labor demand as consisting of factor prices, demand shocks, and lagged employment (Nickell (1984), Bentolila and Saint-Paul (1992), Konings and Roodhooft (1997), Addison and Teixeira (2001)). This strand of research has not, however, rigorously analyzed the effect of uncertainty on labor demand. The process of capital-labor substitution has been

---

4 Competition interactions have been examined in theoretical work by e.g. Grenadier (1999), who shows that the value of the option to delay investment is decreasing in the level of market competition.

examined in empirical work by Ghosal (1991) and Green, Lensink, and Murinde (2001). For a sample of 125 U.S. manufacturing industries, Ghosal (1991) found a significant negative relationship between demand uncertainty and the capital to labor ratio. On the contrary, Green et al. (2001) found for a sample of Polish firms a significant positive relationship between demand uncertainty and the capital to labor ratio.

The purpose of the current paper is to provide empirical evidence regarding the effect of uncertainty on firm investment behavior utilizing Finnish panel data covering the time period 1987 – 2000. The empirical results reveal a significant negative effect of idiosyncratic uncertainty on capital expenditures. Furthermore, the results reveal a structural shift in this effect between the sub-periods 1987 – 1993 and 1994 – 2000. Regarding firm labor policy, I find a modest negative relationship between uncertainty and labor demand.

The remainder of the paper is organized as follows. Section II presents the empirical specifications of the paper, section III presents the estimation results, and finally, section IV provides a summary and conclusions.

II. Empirical Specifications

A. Measuring Uncertainty

As the task of the paper is to examine the impact of uncertainty on firm behavior, it is evident that the measurement of uncertainty plays a crucial role in the empirical analysis. In line with existing empirical work, this paper will utilize the volatility of the firm's stock returns as a proxy variable for uncertainty. Bulan (2001) argues that the advantage of this measure is the fact that it captures the total uncertainty relevant to the firm in a single variable.[5] Furthermore, Bloom et al. (2001) argue that the use of a stock return based measure of uncertainty provides a forward looking proxy for the uncertainty of the firm's environment, implicitly weighted in accordance with the impact of these variables on profits. An important aspect to consider is the effect of leverage on stock return volatility. Leahy and Whited (1996) argue that, *ceteris paribus*, the volatility of the firm's stock returns will increase with the financial leverage of the

---

5 More specifically, Bulan (2001) argues that all sources of uncertainty in the firm's environment such as output price, costs, as well as macro-variables are reflected in the volatility of the firm's stock returns.

firm. Hence, this paper will employ a volatility measure adjusted by the firm's financial leverage.[6]

According to models of irreversible investment, the relevant source of uncertainty for the firm is the total uncertainty reflecting the future evolution of the underlying state variables. On the other hand, CAPM-based models imply that only systematic sources of risk should be relevant for firms' capital investment decisions. Thus, in this setting it seems appropriate to decompose the total uncertainty of the firm, i.e., the volatility of the stock returns, into systematic and idiosyncratic components. Following Bulan (2001), I decompose the total uncertainty of the firm into systematic and idiosyncratic components by estimating the following single index-model of returns for firm i in year t:

$$r_{it} = \alpha_{it} + \beta_{it} r_{mt} + \varepsilon_{it} , \qquad (1)$$

where r_{it} denotes the return of firm i on day t, and where r_{mt} has the same interpretation for the return on the market index.[7] The estimated parameter β corresponds to the CAPM-β, and thus, in order to specify the relevant systematic risk measure for the firm I multiply the estimated β coefficients with the annualized volatility of market returns. As was the case for the estimated volatility of the firm, the estimated CAPM-β coefficients will be affected by financial leverage. Thus, in order to control for financial leverage, I multiply the estimated β coefficients by the market equity ratio of the firm. Furthermore, the single-index model in equation (1) implies that all idiosyncratic components of uncertainty will be captured by the error term ε_{it}. Thus, an annual measure of idiosyncratic uncertainty can be specified as follows:

---

6 The effect of financial leverage is controlled for by adjusting the estimated uncertainty measures with the market equity ratio of the firm. This procedure relies on the assumption that the firm's debt is risk-free, which obviously is not a flawless assumption. However, in the absence of superior alternatives, the procedure seems motivated.

7 The market index employed in this setting corresponds to the Helsinki Stock Exchange (HEX) Portfolio total return index. This index is a value-weighted index where all companies traded on the main list of HEX are represented. However, the weight of any individual company is limited to 10%, thus eliminating the dominance of a few extremely large firms traded on HEX (e.g. Nokia). This index is, however, only available for the time period 1991 – 2000. The market index employed during the years 1987 – 1990 corresponds to the WI-index calculated at the Swedish School of Economics and Business Administration. Knif (1988) argues that the WI-index describes the Finnish market portfolio and is well suited for the estimation of beta coefficients.

$$\sigma_{\varepsilon it} = \sqrt{\sum_{t=1}^{n} \varepsilon_{it}^2} \ ,$$ (2)

which implies that the annual measure of idiosyncratic uncertainty corresponds to the square root of the sum of squared residuals for firm i in year t, when the returns have been filtered through the single-index model in equation (1). The aforementioned procedures result in the following proxy variables for uncertainty: σ_{it}, which is the unlevered measure of total uncertainty, $\beta_{it}\sigma_{mt}$, which corresponds to the unlevered measure of systematic uncertainty, and $\sigma_{\varepsilon it}$, which represents a measure of idiosyncratic uncertainty.

B. Methodology and Empirical Models

The problem at hand calls for estimating the effect of uncertainty on firm-level capital expenditures and labor demand, where realized values of volatility are used as proxy variables for expected uncertainty. The chosen research design exhibits characteristics of endogeneity, which implies that the employed regressors are correlated with the error term. In this case, it seems motivated to employ a Generalized Method of Moments (GMM) estimation procedure, which eliminates the problem caused by endogenous regressors. Furthermore, the empirical phenomena under investigation can be regarded as dynamic in nature, which implies that the inclusion of lagged dependent variables in the empirical specifications seems appropriate. Thus, the statistical model to be utilized in the empirical examination can be specified as:

$$y_{it} = \sum_{k=1}^{p} \alpha_k y_{i(t-k)} + \beta'(L)x_{it} + \lambda_t + \eta_i + v_{it} \ ,$$ (3)

$$t = q+1,\ldots,T_i \ ,$$

$$i = 1,\ldots,N \ .$$

Where η_i and λ_t are individual and time specific effects, x_{it} is a vector of explanatory variables, $\beta(L)$ is a vector of associated polynomials in the lag operator and q is the maximum lag length in the model. The statistical model is the general specification of the dynamic panel data estimators.[8]

The empirical estimation is initiated by constructing benchmark specifications for capital expenditures and labor demand. I use a standard Q theory model of investment as a benchmark to control for the firm's investment opportunities. The Q theory relates the rate of the firm's investment (investment to capital ratio) to its marginal Q, i.e., the present value of all future marginal returns to capital. Marginal Q is in practice measured by Tobin's Q (average Q), which will be utilized in the empirical specification. Further variables that in previous studies have been found to successfully explain firm investment are related to the firm's output (e.g. Leahy and Whited (1996), Henley et al. (2000), Bloom et al. (2001), Bulan (2001)). Thus, I will employ the firm's output to capital ratio as an additional control variable for the firm's investment rate. Finally, in order to control for dynamic adjustment, a lagged dependent variable is incorporated in the empirical specification. Hence, the benchmark model explaining the firm's investment rate can be defined as follows:

$$\left[\frac{I}{K}\right]_{it} = \alpha_1\left[\frac{I}{K}\right]_{i,t-1} + \alpha_2\left[\frac{Y}{K}\right]_{i,t-1} + \alpha_3 Q_{i,t-1} + \lambda_t + \eta_i + v_{it}. \qquad (4)$$

Specifications of the variables are presented in Appendix A. All macroeconomic events specific to a given year are captured by λ_t, whereas all unobservable variables specific to the individual firm are captured by the time-invariant component η_i. The error term is denoted by v_{it}. The benchmark model of capital expenditures is extended by introducing the proxy variables for uncertainty discussed above. The first specification includes the unlevered measure of total uncertainty, σ_{it}. The second specification incorporates the measures $\beta_{it}\sigma_{mt}$ and $\sigma_{\varepsilon it}$, i.e., the unlevered measure of systematic uncertainty and idiosyncratic uncertainty.

The benchmark model for labor demand is constructed on the basis of previous empirical work examining labor dynamics (Nickell (1984), Bentolila and Saint-Paul

---

[8] See Arellano and Bond (1991), Arellano and Bover (1995), and Blundell and Bond (1998) for a review of dynamic panel data estimators.

(1992), Konings and Roodhooft (1997), Addison and Teixeira (2001)). These types of models explain labor demand on the basis of factor prices, demand shocks, and lagged employment. The benchmark model of labor demand utilized in the current paper is specified as follows:

$$L_{it} = \alpha_1 L_{i,t-1} + \alpha_2 \left[\frac{W}{L} \right]_{it} + \alpha_3 \left[\frac{W}{L} \right]_{i,t-1} + \alpha_4 K_{it} + \alpha_5 DS_{it} + \lambda_t + \eta_i + v_{it} . \tag{5}$$

This specification implies that the desired employment, L of firm i in period t (labor demand) is dependent on lagged employment, unit costs of labor and capital, and demand shocks. Employment is measured as the logarithm of labor. The unit cost of labor is obtained by dividing total wages by total employment, whereas the unit cost of capital is approximated by the logarithm of the capital stock, K.[9] The demand shock is measured as the logarithmic change in output. The coefficients λ_t and η_i have the same interpretations as in the benchmark investment model. In a similar manner as for capital expenditures, the labor demand model is extended to incorporate the effects of uncertainty, i.e., by initially including a measure of total uncertainty, and secondly, by including measures of systematic and idiosyncratic uncertainty.

The empirical investigation in the current paper covers the time period 1987 – 2000. Several significant features have influenced the Finnish economy during this particular time period, e.g., a severe economic recession during the years 1990 – 1994, and the seizure of restricting foreign ownership in Finnish publicly traded firms in 1993. Furthermore, incentive schemes linking managerial compensation to firm performance has increased markedly in the latter part of the 1990s. Thus, one might suspect that abnormal economic conditions on the one hand, and the introduction of a new era regarding corporate governance might influence firm behavioral dynamics. Hence, the empirical specifications are further extended by interacting all uncertainty measures with a dummy variable that takes the value of 0 if the period corresponds to the sub-period 1987 – 1993, and 1 if the observation corresponds to the period 1994 – 2000.

---

[9] The unit cost of labor W/L is included as current and as a one-year lag. This specification was found to be suitable and is in line with existing empirical work, e.g., Addison and Teixeira (2001).

This procedure allows for the analysis of whether there has been a structural shift in the effect of uncertainty.[10]

C. Data Sources and Sample Selection

The employed data in the current paper consists of firm-level panel data collected by the Research Institute of the Finnish Economy ETLA. More specifically, the utilized data corresponds to the ETLA 2001-file for Finnish publicly traded firms on the main list of HEX. The total sample covers the time period 1987 – 2000. The minimum requirement is set at three annual observations, which implies that the panel is unbalanced. The stock return data is obtained from two sources. For firms traded on the main list of HEX, data is obtained from the database of the Swedish School of Economics and Business Administration, and if firms have prior listings on the other lists of HEX, the stock return data is obtained from DataStream. The data sources are compatible, both consisting of firm total stock returns. A minimum requirement of 60 daily stock returns is chosen as the restriction for inclusion of annual firm observations.

Banks and insurance companies are excluded from the sample. In order to control for mergers and large restructurings, observations with an absolute change in book value of assets exceeding 30% are excluded from the sample. All uncertainty measures are estimated according to firm accounting periods. Furthermore, an accounting period range of 10 to 14 months is set as a restriction for valid annual firm observations. The aforementioned restrictions result in a sample of 57 firms, with a total of 450 annual observations.

---

[10] It should be noticed that the basic specifications incorporate time fixed effects in order to capture any macroeconomic events specific to any given year. However, in the extended case we are considering differences in uncertainty effects between the sub-periods 1987 – 1993 and 1994 – 2000.

III. Estimation Results

Table I provides descriptive statistics for the sample. Inspection of table I reveals that investment to capital ratios exhibit a relatively high degree of dispersion, with a mean value of 0.252, and minimum and maximum values of 0.005 and 0.827, respectively. Firm labor growth shows a mean value of 0.020, with corresponding minimum and maximum values of –0.419 and 0.855. Furthermore, the figures describing labor- and capital stock reveal substantial differences in magnitude between firms. Regarding the proxy variables for uncertainty, the mean values for total, systematic, and idiosyncratic uncertainty correspond to 0.175, 0.052, and 0.373, respectively.

Table I
Descriptive Statistics of the Variables

Descriptive statistics for the employed variables in the sample. The sample covers the time period 1987 – 2000, and consists of 57 Finnish firms. Labor stock is defined in physical units, whereas the capital stock is specified in monetary units (mFIM). Total uncertainty is specified as the unlevered annualized standard deviation of firm stock returns. Systematic uncertainty corresponds to the unlevered beta coefficient of the firm multiplied with the annualized standard deviation of market returns. Idiosyncratic uncertainty is estimated as the square root of the sum of squared residuals, when firm stock returns have been filtered through a market-index model.

Variable		Mean	Median	Std. Dev.	Minimum	Maximum	Obs.
Investment to capital	I/K	0.252	0.223	0.150	0.005	0.827	450
Output to capital	Y/K	3.404	2.561	2.759	0.593	20.304	450
Tobin's Q	Q	3.831	2.646	3.365	0.726	25.294	450
Labor stock	L	6 657	3 980	7 442	219	41 800	450
Capital stock	K	2 801	1 243	5 324	15	50 197	450
Labor growth	ΔL	0.020	0.016	0.143	-0.419	0.855	450
Wages to labor	W/L	0.185	0.182	0.049	0.070	0.342	450
Output growth	ΔY	0.077	0.068	0.135	-0.389	0.851	450
Total uncertainty	σ_{it}	0.175	0.155	0.098	0.022	0.711	450
Systematic uncertainty	$\beta_{it}\sigma_{mt}$	0.052	0.043	0.049	-0.025	0.387	450
Idiosyncratic uncertainty	$\sigma_{\dot{a}t}$	0.373	0.337	0.154	0.128	1.129	450

Table II presents the results from the initial part of the empirical analysis, i.e., the effect of uncertainty on capital expenditures. Estimation uses DPD for Ox (see Doornik, Arellano, and Bond (2001)). All GMM estimations are performed in first-differenced form in order to eliminate firm fixed effects. Transformation, taking lags, and instrumentation reduce the original sample of 450 annual observations to 305 observations. The table reports estimation results for models (I) – (V), where model (I) corresponds to the benchmark Q theory model of investment, excluding proxy variables for uncertainty. Models (II) and (IV) correspond to specifications incorporating the effects of total uncertainty, and systematic and idiosyncratic uncertainty, respectively. Finally, models (III) and (V) are extended specifications, where the uncertainty variables have been interacted with multiplicative dummy variables allowing for distinction between the sub-periods 1987 – 1993 and 1994 – 2000. The chosen GMM estimation procedure allows all lagged values of regressors dated t-s for $s \geq 2$ to be used as valid instruments. However, using the whole history of the series as instruments has often been found to be problematic in small samples, and may result in over-fitting biases. Thus, in the current setting, the instrument set in all specifications consists of the second lagged level up to a maximum of the seventh lagged level of all right-hand side (RHS) variables, and additionally of the logarithm of total assets. Inspection of the diagnostic tests reveal that the chosen models seem to perform well, in all specifications the Wald test statistic (joint) shows that the null hypothesis that the chosen independent variables have no effect on the dependent variable is safely rejected. This is also the case for the Wald test statistic regarding the chosen dummy variables, where the null hypothesis of no association, likewise, is safely rejected. Furthermore, the Sargan test of instrument validity cannot reject the null hypothesis that the chosen instrument set is valid and the models are correctly specified. Finally, the AR (2) test for second order autocorrelation shows that the null hypothesis of lacking second order autocorrelation cannot be rejected in any of the specifications.[11]

---

[11] The lack of second order autocorrelation in the residuals is essential for the GMM-estimator to be consistent (Doornik, Arellano, and Bond (2001)).

Table II
Estimation Results for Capital Expenditures

The table reports 1-step estimation results using robust standard errors, two-tailed p-values in parentheses. All specifications include a full set of time dummies. The diagnostic tests include the Wald statistic (joint), where the null hypothesis states that the chosen independent variables have no effect on the dependent variable. The null hypothesis for the Wald statistic (dummy) is that the chosen dummy variables have no explanatory power in the estimation. The Sargan test statistic is employed in order to analyze the validity of the chosen instrument set. The null hypothesis for the Sargan test is that the instrument set is valid, and that the model is correctly specified. AR (2) is a test for second order autocorrelation. The null hypothesis states no autocorrelation for the error term, in which case the test statistic is asymptotically normally distributed. The uncertainty variables σ_{it}, $\beta_{it}\sigma_{mt}$, and $\sigma_{\hat{e}it}$ are interacted in models (III) and (V) with a dummy variable D, which takes on a value of 0 if the observation corresponds to the period 1987 – 1993, and 1 if the observation corresponds to the period 1994 – 2000.

Dependent variable	Independent variables									
$\Delta\left(\frac{I}{K}\right)_{it}$	$\Delta\left(\frac{I}{K}\right)_{i,t-1}$	$\Delta\left(\frac{I}{K}\right)_{i,t-1}$	$\Delta Q_{i,t-1}$	$\Delta\sigma_{it}$	$\Delta\beta_{it}\sigma_{mt}$	$\Delta\sigma_{\hat{e}it}$	$\Delta D\sigma_{it}$	$\Delta D\beta_{it}\sigma_{mt}$	$\Delta D\sigma_{\hat{e}it}$	
I	0.166***	0.018**	0.014**							
	(0.004)	(0.044)	(0.012)							
II	0.191***	0.016**	0.017**	-0.037						
	(0.001)	(0.036)	(0.012)	(0.751)						
III	0.191***	0.016**	0.017**	-0.064			0.034			
	(0.001)	(0.038)	(0.013)	(0.722)			(0.847)			
IV	0.176***	0.018**	0.015**		0.256	-0.126**				
	(0.002)	(0.020)	(0.028)		(0.329)	(0.021)				
V	0.152**	0.018**	0.014**		0.031	-0.080		0.359	-0.158**	
	(0.010)	(0.011)	(0.038)		(0.954)	(0.214)		(0.497)	(0.012)	

Diagnostic tests	Wald (joint)	Wald (dummy)	Sargan	AR (2)	Obs.
I	25.570***	41.230***	249.000	-0.962	305
	(0.000)	(0.000)	(0.130)	(0.336)	
II	42.470***	39.920***	292.400	-1.009	305
	(0.000)	(0.000)	(0.307)	(0.313)	
III	44.450***	36.640***	290.800	-1.008	305
	(0.000)	(0.000)	(0.316)	(0.314)	
IV	45.020***	52.670***	314.500	-1.045	305
	(0.000)	(0.000)	(0.805)	(0.296)	
V	59.560***	43.820***	311.200	-1.095	305
	(0.000)	(0.000)	(0.820)	(0.274)	

Inspection of the parameter estimates of the benchmark specification (I) reveals that all coefficient estimates show the expected signs and are significant on a minimum of 5%. Thus, in line with previous research, it seems as if a Q theory model seems to explain firms' investment rates quite efficiently. Furthermore, the dynamic specification incorporating the lagged dependent variable seems to be appropriate, which has been proven valid in numerous previous studies. Models (II) and (III) incorporate the effect of total uncertainty on capital expenditures. The parameter estimate for total uncertainty in model (II) is negative, however, not significant on any conventional levels. Furthermore, the estimate of the interaction dummy variable in model (III) does not reveal any signs of a structural shift regarding this effect. Models (IV) and (V) include the effect of systematic and idiosyncratic uncertainty. Somewhat surprisingly, the coefficient estimate of systematic uncertainty is positive, although not significant. On the contrary, the coefficient of idiosyncratic uncertainty is negative, and statistically significant on a 5% level. Examination of the results from specification (V) provides more insight to this result, i.e., the coefficient estimate corresponding to the sub-period 1987 – 1993 is insignificantly different from zero. However, the interaction dummy variable is negative and significant on a 5% level. Hence, the results imply a structural shift regarding this effect. Furthermore, the deduced coefficient of idiosyncratic uncertainty for the sub-period 1994 – 2000 corresponds to –0.238 (–0.080–0.158), which is statistically significant on a 1% level.[12]

Table III presents the results for labor demand. The table reports estimation results for specifications (I) – (V), which are constructed in an identical manner as for capital expenditures. The instrument set in all specifications consists of the second lagged level up to a maximum of the third lagged level of all RHS variables. Inspection of the diagnostic tests reveal a similar picture as in table II, i.e., the Wald test statistics (joint and dummy) safely reject the null hypotheses of lacking association between the dependent and independent variables, and invalid dummy variables, respectively. Furthermore, the Sargan and AR (2) tests cannot reject the null hypotheses of instrument validity, and lack of second order autocorrelation in residuals, respectively. Inspection of the benchmark specification, model (I), reveals that the estimated coefficients are in

---

[12] The t-statistic for the linear restriction test under the null hypothesis that the coefficient estimate for the sub-period 1987 – 1993 added with the estimated difference in slope coefficients is zero, corresponds to – 3.99.

line with previous empirical work regarding labor demand. It is interesting to notice that the coefficient estimates for the unit cost of labor are highly significant both as a current and a one-year lag. Furthermore, the proxy variable for the unit cost of capital is positive and significant on a 5% level. Finally, the coefficient estimate for demand shock is positive and highly significant. Models (II) and (III) extend the benchmark specification by incorporating the effect of total uncertainty. The estimated coefficient of total uncertainty in specification (II) is negative and significant on a 10% level. Inspection of specification (III) implies that there is no statistically significant structural shift regarding this effect between the two sub-periods. Specification (IV) incorporates the effects of systematic and idiosyncratic uncertainty. Both coefficient estimates are negative, however, not significant on any conventional levels. Finally, examination of the estimated coefficients of specification (V) reveals no evidence of a structural shift regarding the effects of systematic and idiosyncratic uncertainty between the sub-periods.

Table III
Estimation Results for Labor Demand

The table reports 1-step estimation results using robust standard errors, two-tailed p-values in parentheses. All specifications include a full set of time dummies. The diagnostic tests include the Wald statistic (joint), where the null hypothesis states that the chosen independent variables have no effect on the dependent variable. The null hypothesis for the Wald statistic (dummy) is that the chosen dummy variables have no explanatory power in the estimation. The Sargan test statistic is employed in order to analyze the validity of the chosen instrument set. The null hypothesis for the Sargan test is that the instrument set is valid, and that the model is correctly specified. AR (2) is a test for second order autocorrelation. The null hypothesis states no autocorrelation for the error term, in which case the test statistic is asymptotically normally distributed. The uncertainty variables σ_{it}, $\beta_{it}\sigma_{mt}$, and $\sigma_{\varepsilon it}$ are interacted in models (III) and (V) with a dummy variable D, which takes on a value of 0 if the observation corresponds to the period $1987 - 1993$, and 1 if the observation corresponds to the period $1994 - 2000$.

Dependent variable	Independent variables										
ΔL_{it}	$\Delta L_{i,t-1}$	$\Delta\left(w/L\right)_{it}$	$\Delta\left(w/L\right)_{i,t-1}$	ΔK_{it}	ΔDS_{it}	$\Delta\sigma_{it}$	$\Delta\beta_{it}\sigma_{mt}$	$\Delta\sigma_{\varepsilon it}$	$\Delta D\sigma_{it}$	$\Delta D\beta_{it}\sigma_{mt}$	$\Delta D\sigma_{\varepsilon it}$
I	0.800***	-3.089***	2.724***	0.112**	0.438***						
	(0.000)	(0.000)	(0.001)	(0.048)	(0.000)						
II	0.817***	-3.080***	2.541***	0.119**	0.510***	-0.247*					
	(0.000)	(0.000)	(0.000)	(0.026)	(0.000)	(0.085)					
III	0.814***	-3.030***	2.560***	0.119**	0.503***	-0.344**			0.135		
	(0.000)	(0.000)	(0.000)	(0.025)	(0.000)	(0.049)			(0.447)		
IV	0.801***	-2.946***	2.257***	0.143***	0.438***		-0.437	-0.041			
	(0.000)	(0.000)	(0.001)	(0.006)	(0.000)		(0.246)	(0.419)			
V	0.796***	-2.860***	2.253***	0.145***	0.433***		-0.653	-0.036		0.255	-0.0347
	(0.000)	(0.000)	(0.001)	(0.005)	(0.000)		(0.202)	(0.545)		(0.600)	(0.625)

Diagnostic tests	Wald (joint)	Wald (dummy)	Sargan	AR (2)	Obs.
I	726.300***	124.800***	77.780	1.468	305
	(0.000)	(0.000)	(0.750)	(0.142)	
II	882.700***	111.800***	95.680	1.312	305
	(0.000)	(0.000)	(0.815)	(0.190)	
III	940.100***	86.030***	94.940	1.279	305
	(0.000)	(0.000)	(0.811)	(0.201)	
IV	1165.000***	145.100***	111.700	1.539	305
	(0.000)	(0.000)	(0.887)	(0.124)	
V	1236.000***	87.330***	110.000	1.523	305
	(0.000)	(0.000)	(0.886)	(0.128)	

IV. Summary and Conclusions

This paper analyzes the effect of uncertainty on firm-level capital expenditures and labor demand based on a sample of Finnish firms during the time period 1987 – 2000. The empirical results in the current paper reveal that idiosyncratic uncertainty significantly depresses firm-level capital expenditures. This effect is present after controlling for the firm's investment opportunities by Tobin's Q, the firm's output to capital ratio, and dynamic adjustment in capital expenditures. However, it seems as if this effect is concentrated to the latter part of the investigation period, i.e., 1994 – 2000. The results fail to find any significant relationship between firm-level capital expenditures and the effects of total and systematic uncertainty. In the case of labor demand, I find a modest negative effect of total uncertainty on labor demand. This effect is present after controlling for unit costs of labor and capital, demand shocks, and dynamic labor adjustment of the firm. Systematic and idiosyncratic uncertainty does not affect the desired employment level of the firm.

The finding that idiosyncratic uncertainty tends to reduce firm capital expenditures is interesting both from a theoretical and empirical perspective. The results of Henley et al. (2000) revealed that idiosyncratic uncertainty tends to increase firm investment. This result was explained based on two propositions, namely, the Hartman-Abel effect, and alternatively, on the basis of a dominating put option effect to resell capital in the future.[13] On the other hand, Bulan (2001) found that idiosyncratic uncertainty depresses firm investment, which thus, corroborates the results in the current paper. However, another explanation for the fact that idiosyncratic uncertainty tends to reduce capital expenditures may arise from managerial risk-preferences. Consider the case when a risk-averse manager's compensation is closely tied to the performance of the firm. If a substantial part of the manager's wealth is linked to the equity of the firm, then this implies that the manager is exposed to idiosyncratic risk. Hence, this might imply that the manager will require a higher rate of return on firm investment than what is dictated by capital markets.[14] The implication of this scenario is that the manager might forego profitable investment opportunities that would increase the risk of the firm, and thus, greater idiosyncratic uncertainty would tend to lower investment. This

---

[13] Henley et al. (2000) argue that a dominating put option effect in the case of idiosyncratic uncertainty is consistent with the suggestion that increased idiosyncratic uncertainty does not depress the resale value of capital whereas increased aggregate uncertainty does.
[14] See e.g. Himmelberg, Hubbard, and Love (2001).

argument might to some extent help to explain the empirical results in the current paper, i.e., the fact that the depressing effect of idiosyncratic uncertainty on capital expenditures is concentrated to the latter part of the 1990s (1994 – 2000). As mentioned earlier, the latter part of the 1990s can be characterized as a new era of corporate governance in Finland, with the seizure of restricting foreign ownership in Finnish publicly traded firms, and by the rapid expansion of incentive schemes linking managerial compensation to firm performance. Hence, if one assumes that this evolution has caused managerial wealth to be more exposed to the firm's equity, then the explanation pertaining to managerial risk-preferences, i.e., that greater idiosyncratic depresses firm investment, seems plausible.

The finding that uncertainty reduces labor demand is interesting, albeit the relationship is only of modest strength. Thus, it seems as if firms respond to an increase in uncertainty about the future by reducing the desired level of employment. This finding is obviously an interesting complement to previous studies regarding labor adjustment dynamics. Indeed, this result implies that the effect of uncertainty on factor demand should not solely be examined from a single factor perspective. Finally, evidence regarding behavioral dynamics of micro-economic entities has clear importance from a macro-economic policy perspective.

References

Abel, A.B. (1983), Optimal Investment under Uncertainty, *American Economic Review*, 73, pp. 228 – 233.

Abel, A.B., A. Dixit, J. Eberly, and R. Pindyck (1996), Options, the Value of Capital, and Investment, *Quarterly Journal of Economics*, 111, pp. 753 – 777.

Addison, J.T., and P. Teixeira (2001), *Employment Adjustment in Portugal: Evidence from Aggregate and Firm Data*, Discussion Paper No. 391, Institute for the Study of Labor (IZA).

Akerlof, G. (1970), The Market for Lemons: Quality Uncertainty and the Market Mechanism, *Quarterly Journal of Economics*, 89, pp. 488 – 500.

Arellano, M., and S.R. Bond (1991), Some Tests of Specification for Panel Data: Monte Carlo Evidence and an Application to Employment Equations, *Review of Economic Studies*, 58, pp. 277 – 297.

Arellano, M., and O. Bover (1995), Another Look at the Instrumental Variables Estimation of Error-Components Models, *Journal of Econometrics*, 68, pp. 29 – 51.

Bentolila, S., and G. Saint-Paul (1992), The Macroeconomic Impact of Flexible Labour Contracts, with an Application to Spain, *European Economic Review*, 36, pp. 1013 – 1053.

Bloom, N., S. Bond, and J. Van Reenen (2001), *The Dynamics of Investment under Uncertainty*, Working Paper, University College London and Institute for Fiscal Studies.

Bo, H., and E. Sterken (2000), *Do Firms Wait to Invest? An Empirical Investigation*, Working Paper, SOM-theme C.

Blundell, R.W., and S.R. Bond (1998), Initial Conditions and Moment Restrictions in Dynamic Panel Data Models, *Journal of Econometrics*, 87, pp. 115 – 143.

Bulan, L.T. (2001), *Real Options, Irreversible Investment and Firm Uncertainty: New Evidence from U.S. Firms*, Working Paper, Columbia University.

Dixit, A, and R. Pindyck (1994), *Investment under Uncertainty*, Princeton, New Jersey, Princeton University Press.

Doornik, J.A., M. Arellano, and S. Bond (2001), *Panel Data Estimation using DPD for Ox*, Nuffield College, Oxford.

Ghosal, V. (1991), Demand Uncertainty and the Capital-Labor Ratio: Evidence from the U.S. Manufacturing Sector, *The Review of Economics and Statistics*, Volume 73, Issue 1, pp. 157 – 161.

Green, C.J., R. Lensink, and V. Murinde (2001), *Demand Uncertainty and the Capital-Labour Ratio in Poland*, Working Paper, University of Birmingham.

Grenadier, S. (1999), *Option Exercise Games: The Case of the Equilibrium Investment Strategies of Firms*, Working Paper, Stanford University.

Hartman, R. (1972), The Effects of Price and Cost Uncertainty on Investment, *Journal of Economic Theory*, 5, pp. 258 – 266.

Hartman, R. (1976), Factor Demand with Output Price Uncertainty, *American Economic Review*, 66, pp. 675 – 681.

Henley, A., A. Carruth, and A. Dickerson (2000), *Industry-Wide Versus Idiosyncratic Uncertainty and Investment: British Company Panel Data Evidence*, Working Paper, University of Wales Aberystwyth.

Himmelberg, C.P., G. Hubbard, and I. Love (2001), *Investor Protection, Ownership and Capital Allocation*, Working Paper, Columbia University.

Knif, J. (1988), *Tests for Market Model Instability, An Empirical Comparison of Tests Using Recursive Residuals*, Research Reports 18, Swedish School of Economics and Business Administration, Helsinki.

Konings, J., and F. Roodhooft (1997), How Elastic is the Demand for Labour in Belgian Enterprises? Results from Firm Level Panel Data, 1986 – 1994, *De Economist*, 145, pp. 229 – 241.

Leahy, J., and T.M. Whited (1996), The Effect of Uncertainty on Investment: Some Stylized Facts, *Journal of Money, Credit and Banking*, 28, pp. 64 – 83.

McDonald, R., and D. Siegel (1986), The Value of Waiting to Invest, *Quarterly Journal of Economics*, (November), 101, pp. 707 – 728.

Nickell, S. (1984), An Investigation of the Determinants of Manufacturing Employment in the United Kingdom, *Review of Economic Studies*, 51, pp. 529 – 557.

Oi, W. (1962), Labor as a Quasi-fixed Factor, *Journal of Political Economy*, 70 (6), pp. 538 – 555.

Pindyck, R. (1988), Irreversible Investment, Capacity Choice, and the Valuation of the Firm, *American Economic Review*, (December), 79, pp. 969 – 985.

Whited, T.M. (1992), Debt, Liquidity Constraints, and Corporate Investments: Evidence from Panel Data, *Journal of Finance*, 47 (4), pp. 1425 – 1460.

Appendix A

This section describes the specifications of the employed variables. Firm accounting data is obtained from the Research Institute of the Finnish Economy ETLA 2001-file for Finnish publicly traded firms on the main list of HEX. All variables are measured in nominal terms, except where noted. In the following description, the subscript i is dropped for convenience.

Investment: I_t is reported gross investment in fixed assets during the accounting period.

Output: Y_t is the reported total sales during the accounting period.

Capital stock: K_t is specified as the replacement value of the capital stock, calculated using a perpetual inventory method, see Whited (1992) for details. Starting with the book value of gross plant, property, and equipment for the firm in the first year as a proxy for the replacement value of the capital stock, I estimate K_t as follows:

$$K_t = \left[K_{t-1}\left(\frac{P_t}{P_{t-1}}\right) + I_t \right]\left(1 - \frac{2}{ULC}\right), \tag{A.1}$$

where P_t corresponds to the price deflator for fixed investment (*Kansantalouden Tilinpito*, sarja: Tukkuhintaindeksi, Investointitavarat, 1995 = 100). *ULC* corresponds to the average of ULC_t, i.e., the useful life of capital goods. The second term in (A.1) represents the amount of the capital stock that depreciates in each period, and is based on the assumption that economic depreciation is double declining balance. The useful life of capital goods in any period is calculated as follows:

$$ULC_t = \frac{GK_{t-1} + I_t}{DEPR_t}, \tag{A.2}$$

where GK_{t-1} is the reported value of gross plant, property, and equipment, and where $DEPR_t$ corresponds to capital depreciation.

Tobin's Q: the relationship between the market value of the capital stock and the replacement cost is calculated as follows:

$$Q_t = \frac{E_t + D_t}{K_t},$$

(A.3)

where $E_t + D_t$ represents the market value of the capital stock. This specification relies on the assumption that the market value of D_t is equal to its book value.

Market value of equity: E_t is specified as the total market capitalization of the firm. This data is obtained from KOP Pörssiyhtiöt manuals and Kauppalehti databases.

Total debt: D_t is specified as the reported sum of short- and long-term debt.

Labor stock: L_t corresponds to average number of employees during the firm's accounting period.

Wages: W_t is specified as total labor costs during the firm's accounting period.

Agentum Technologies Inc.

Agent Applications

8.5.2002   Agentum Technologies Inc. - Svante Olofsson   

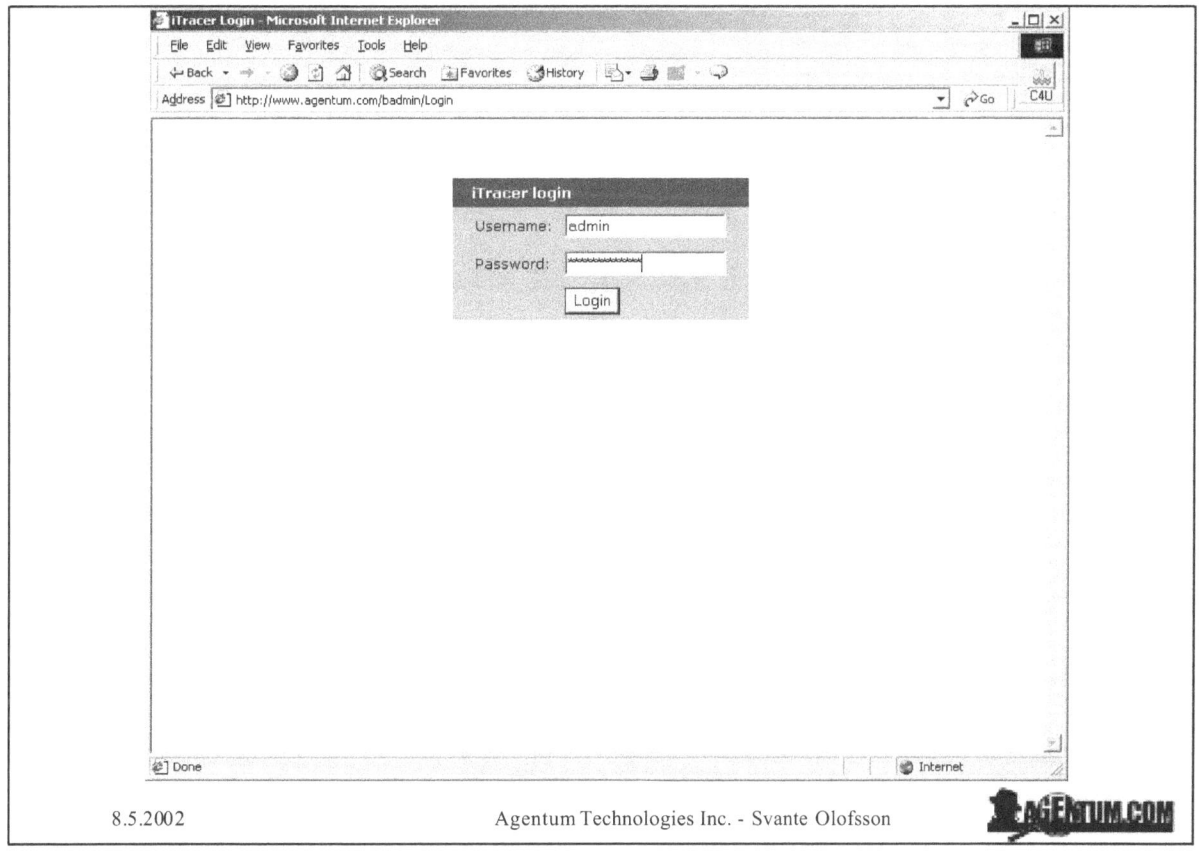

8.5.2002   Agentum Technologies Inc. - Svante Olofsson

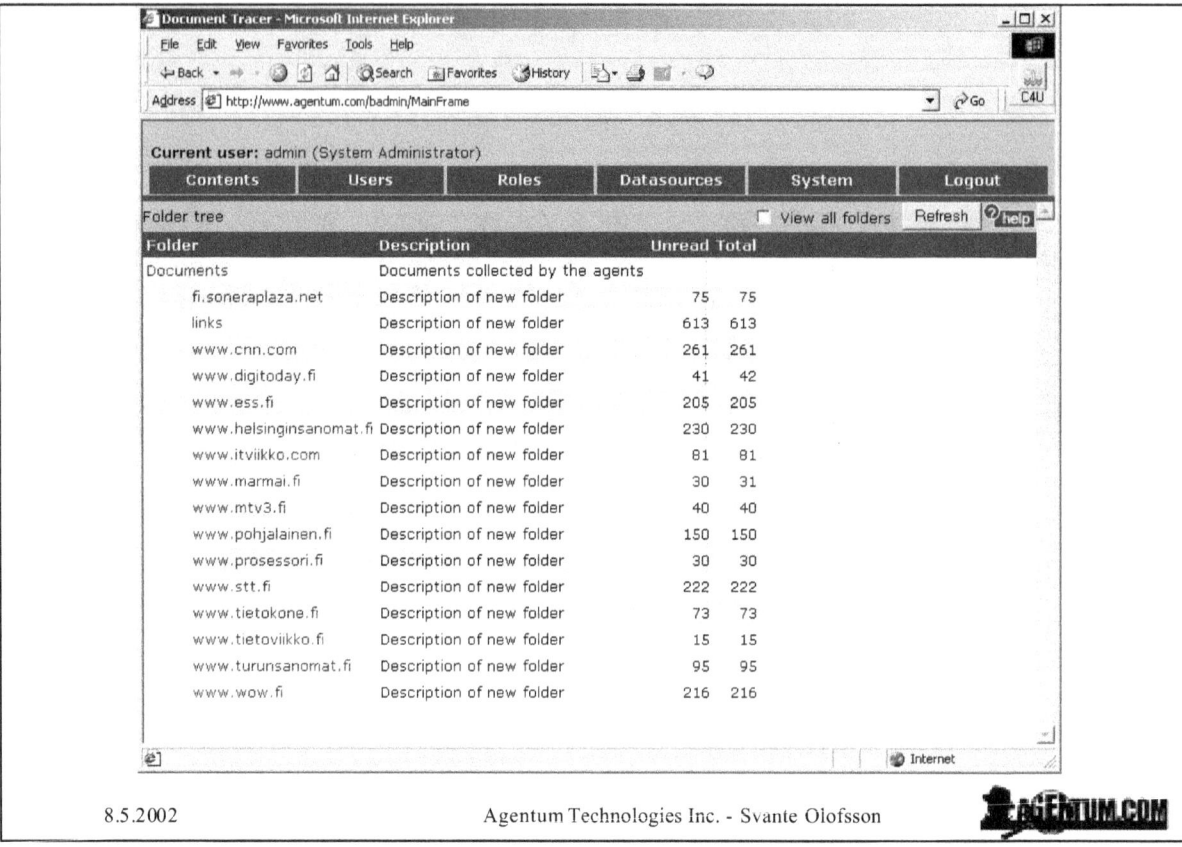

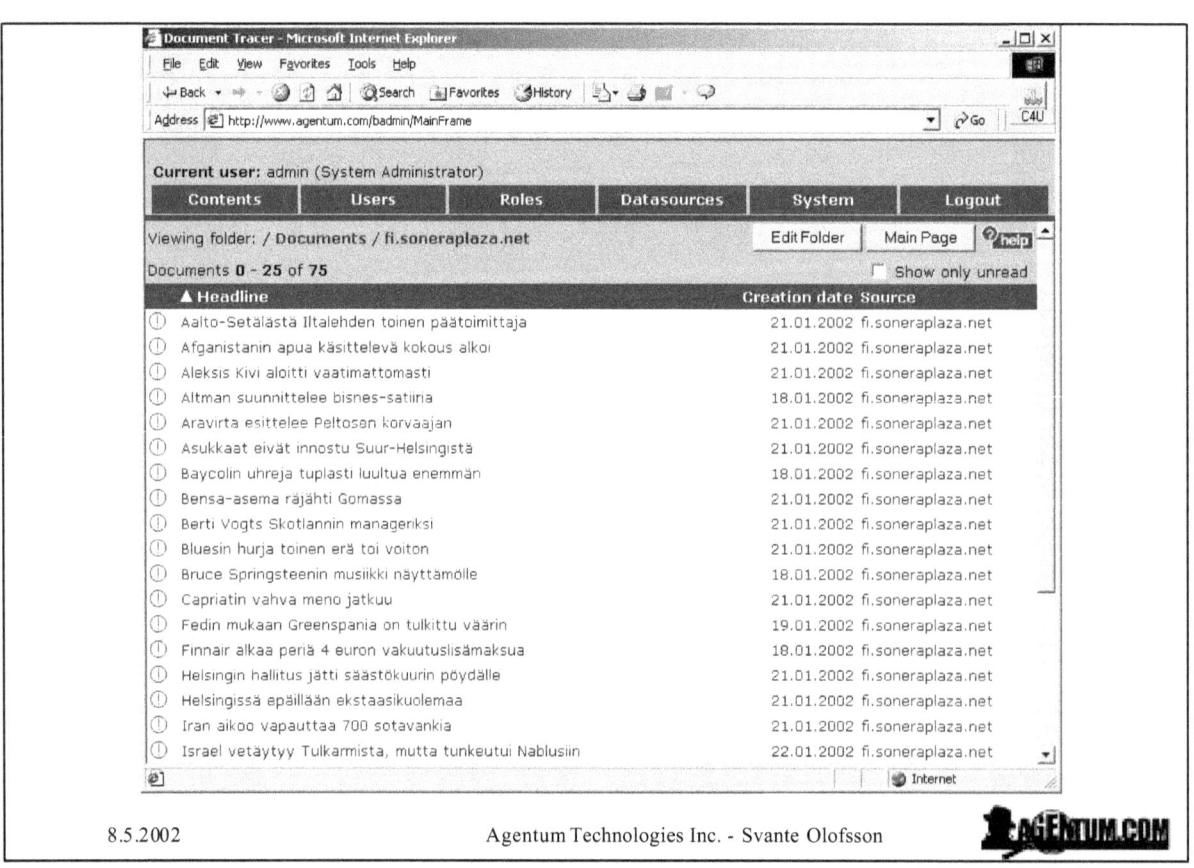

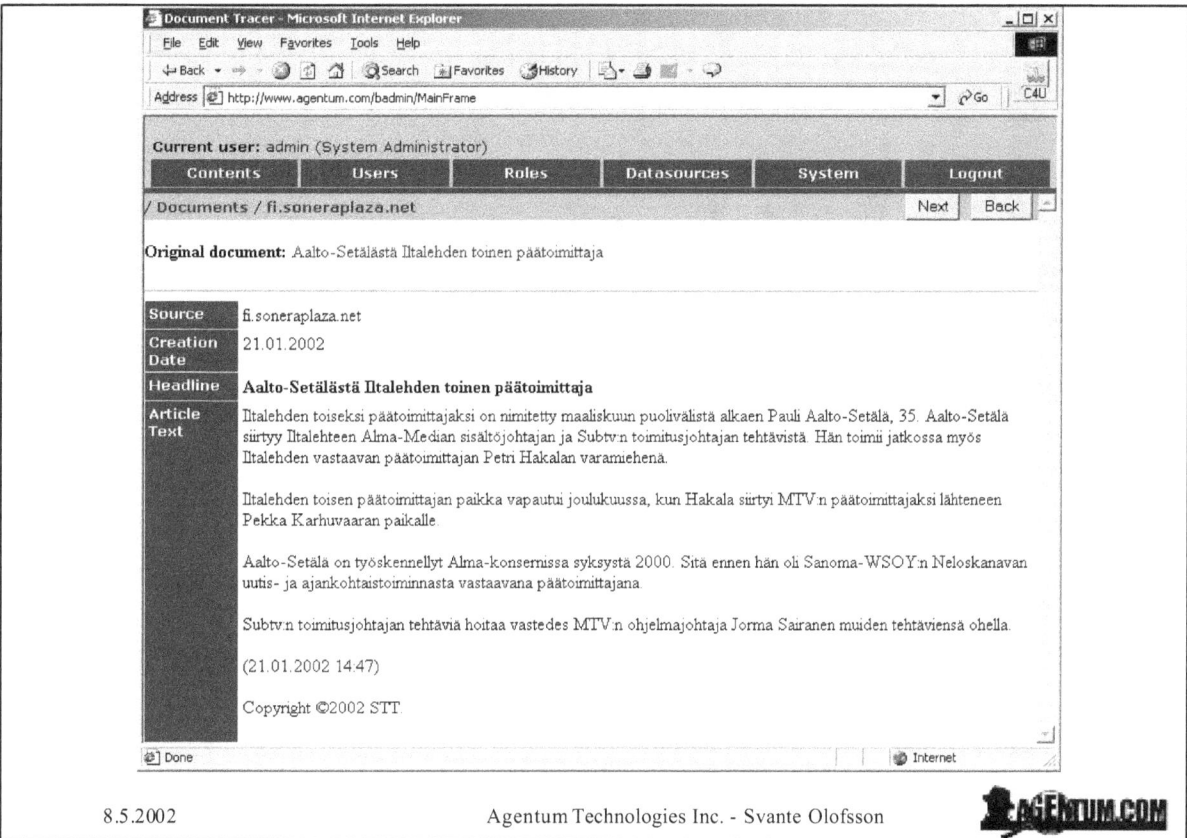

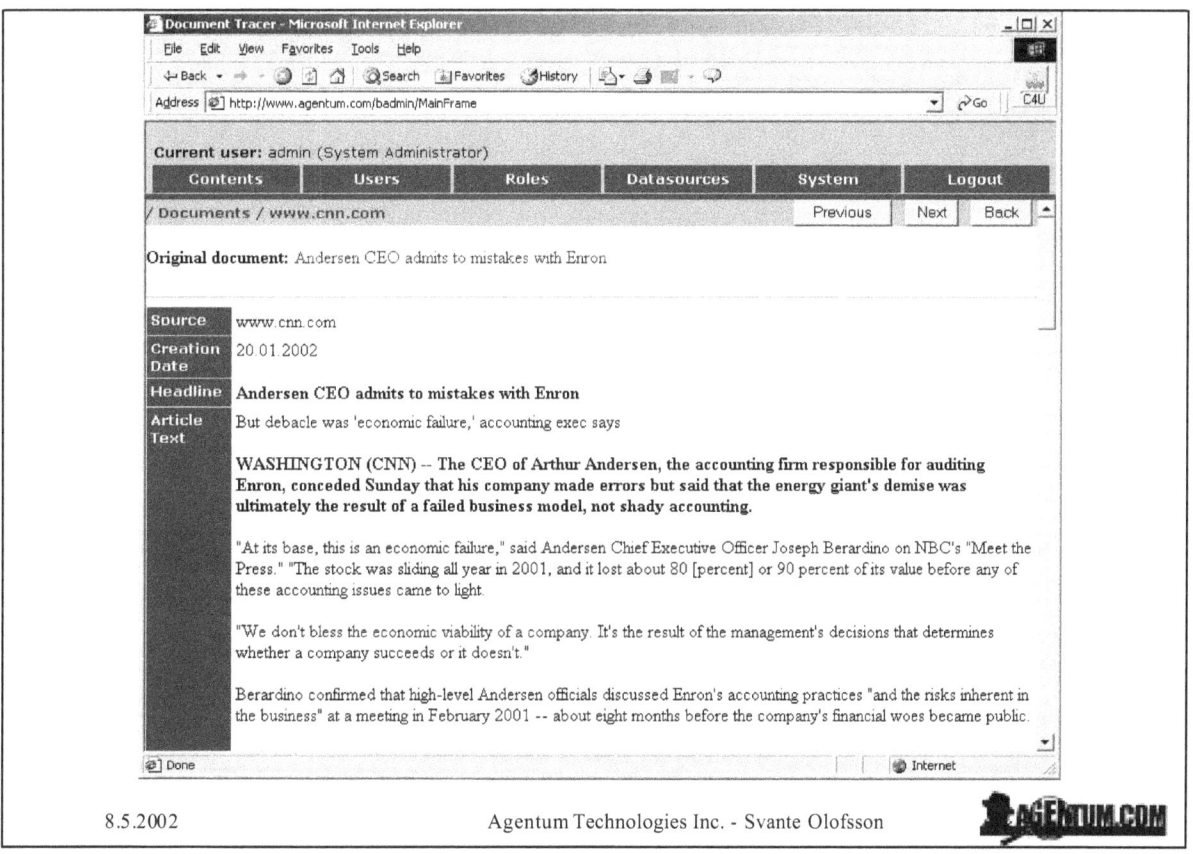

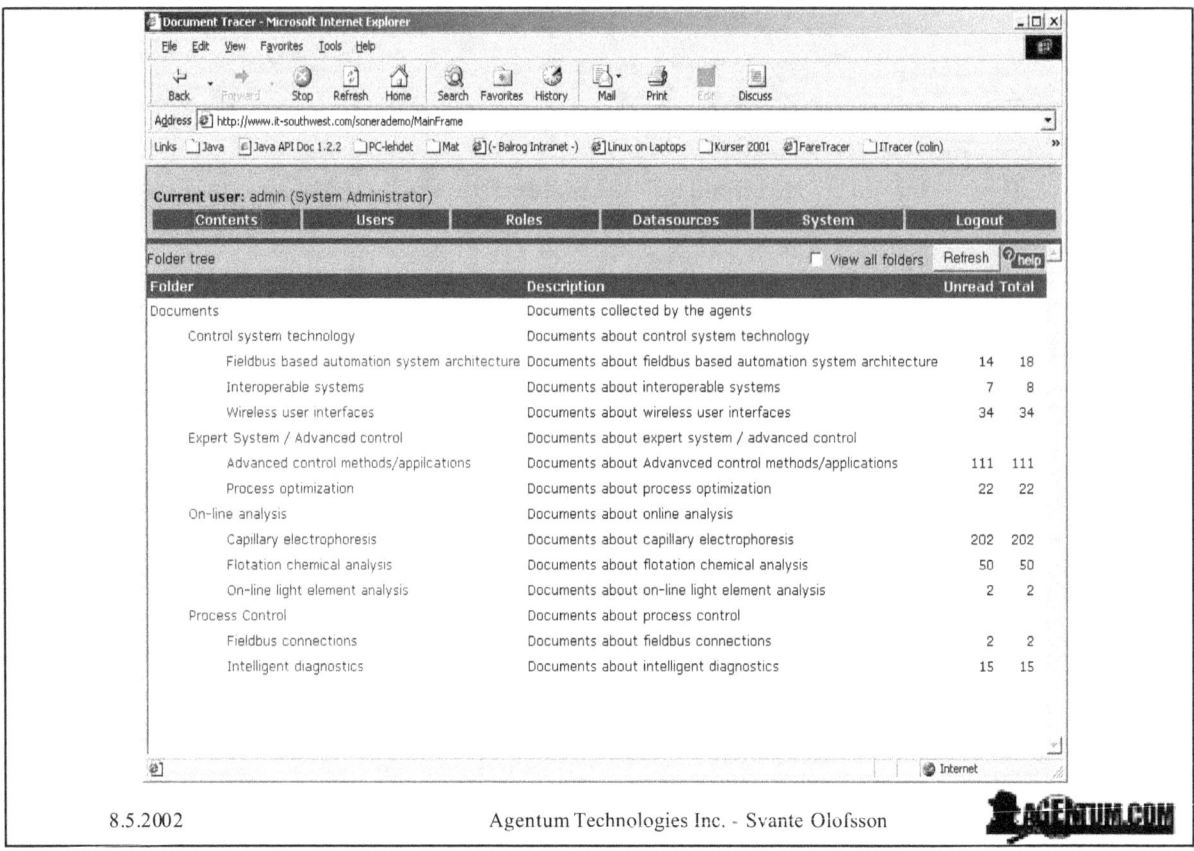

This section contains the 'table of contents' for the storyline.

The complete storyline is built from smaller sections or chapters. The chapters are presented in the table of contents and can be opened in the window to the rigth by clicking on the [View] link.

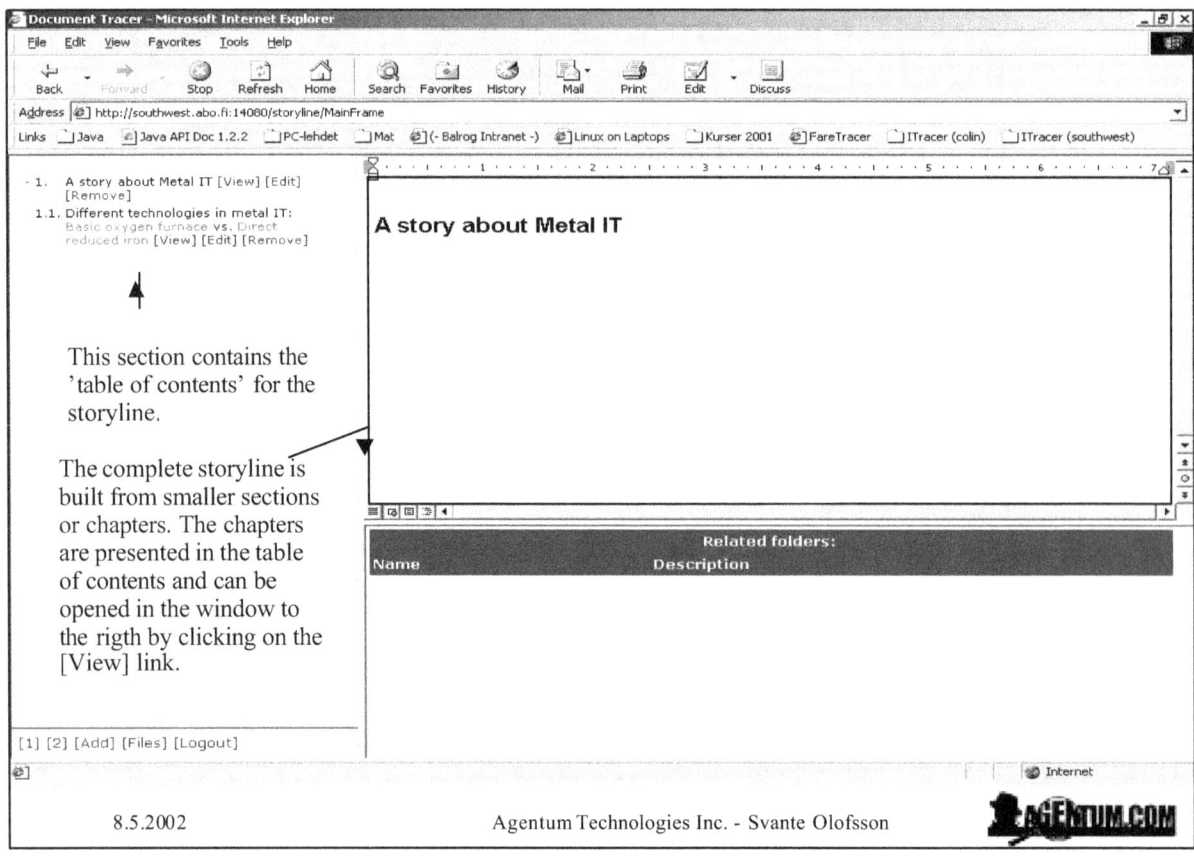

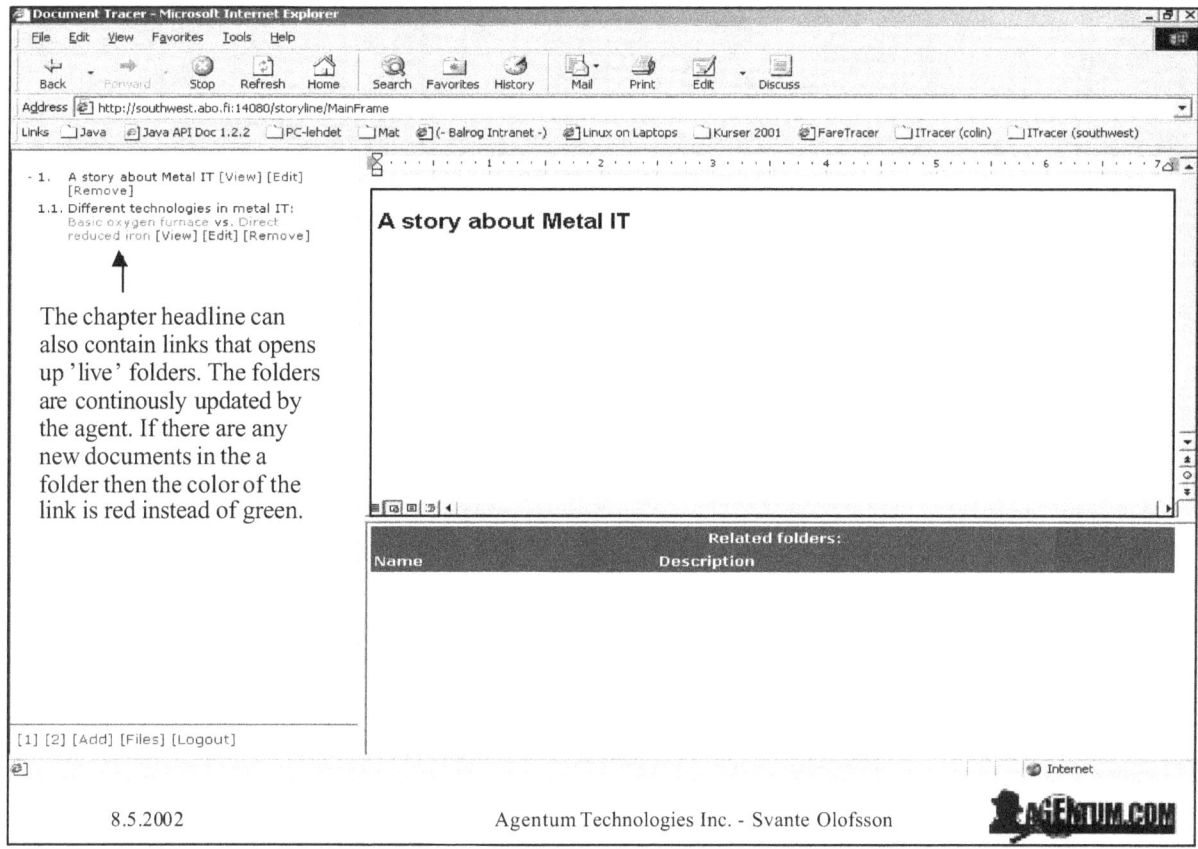

The chapter headline can also contain links that opens up 'live' folders. The folders are continuously updated by the agent. If there are any new documents in the a folder then the color of the link is red instead of green.

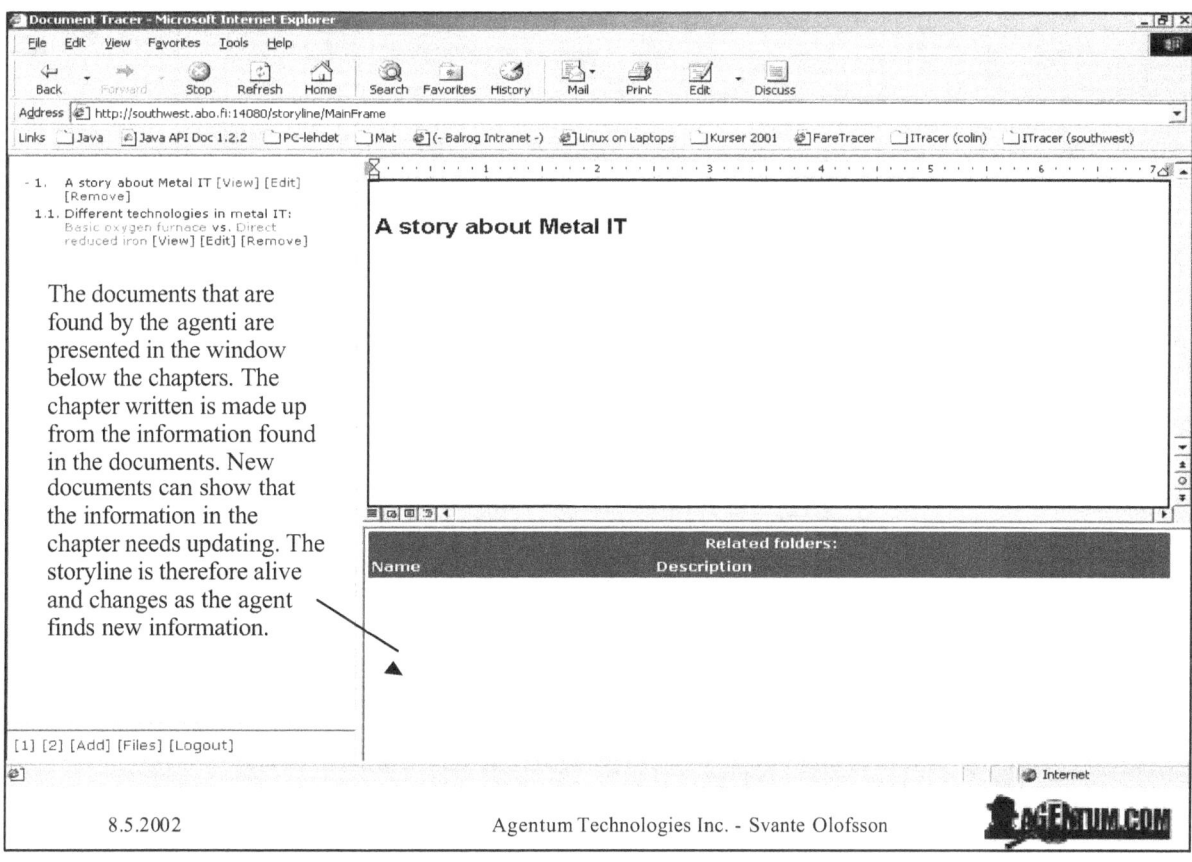

The documents that are found by the agenti are presented in the window below the chapters. The chapter written is made up from the information found in the documents. New documents can show that the information in the chapter needs updating. The storyline is therefore alive and changes as the agent finds new information.

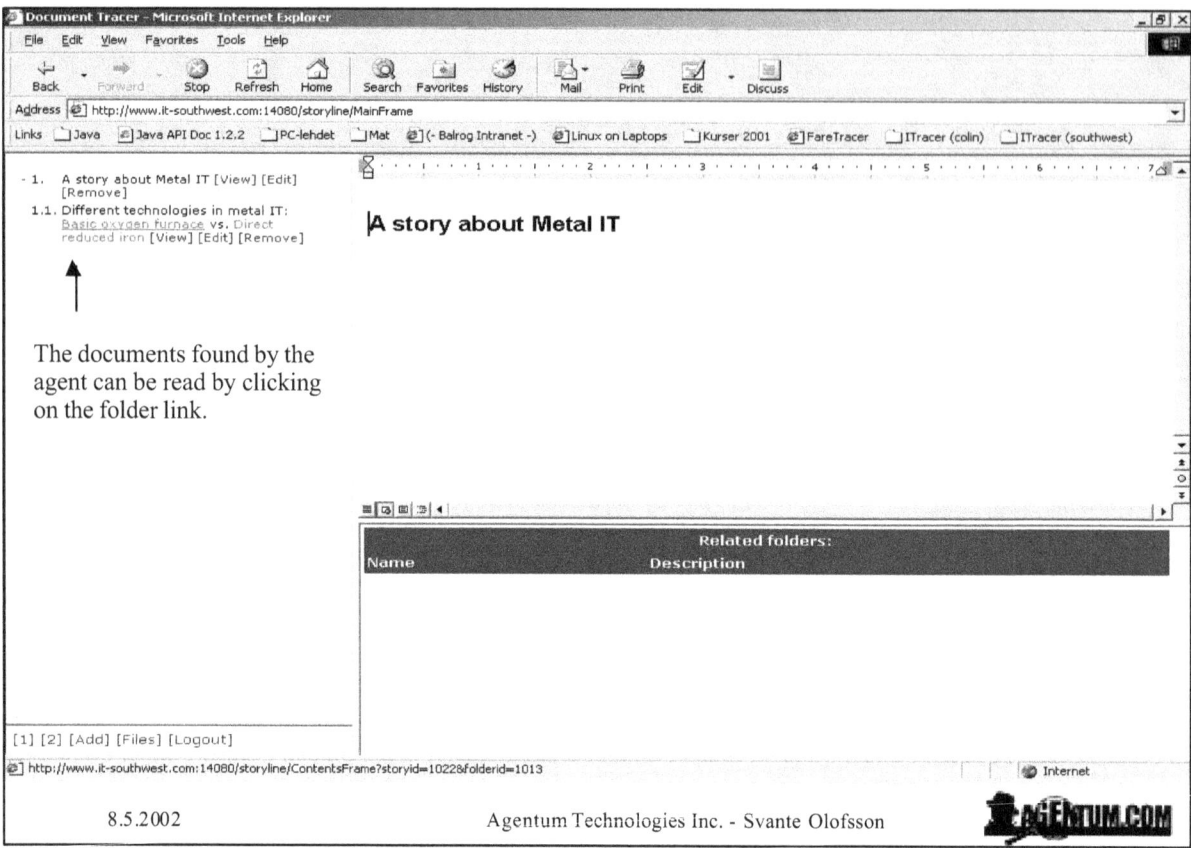

The documents found by the agent can be read by clicking on the folder link.

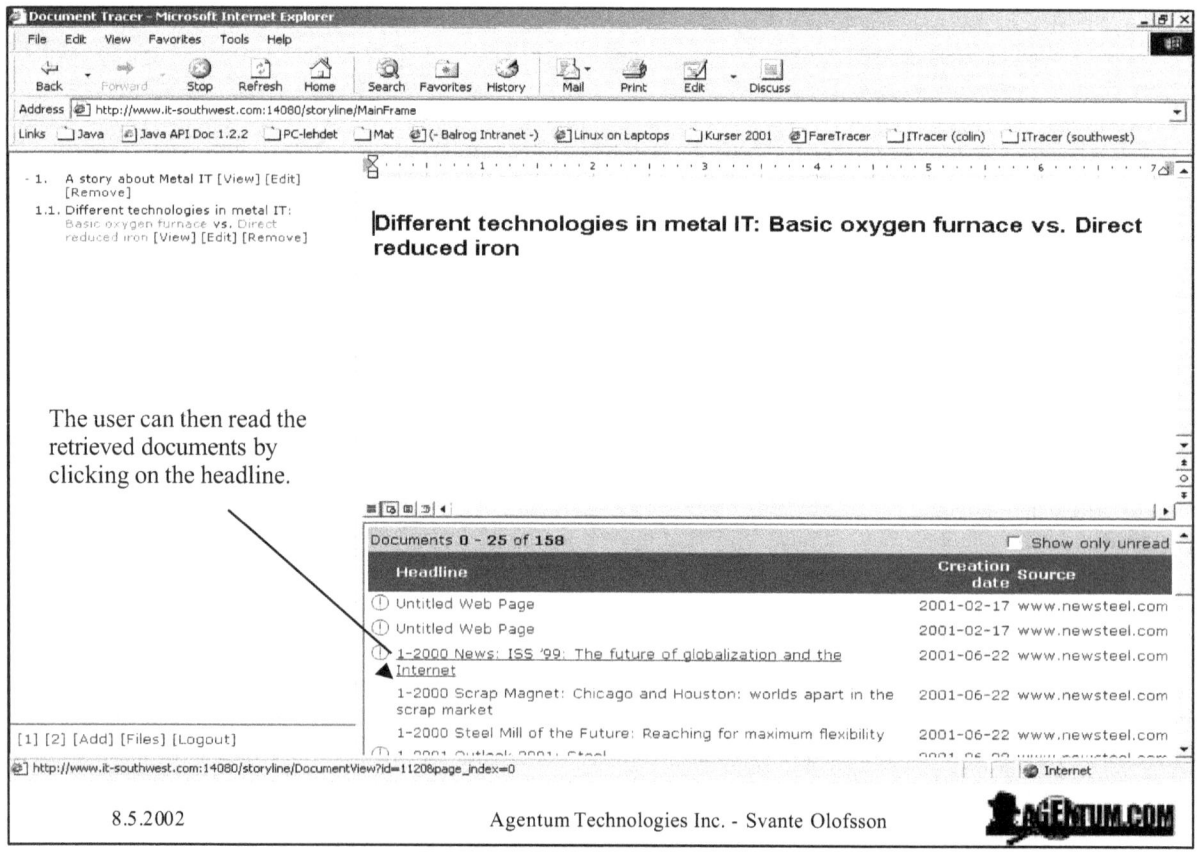

The user can then read the retrieved documents by clicking on the headline.

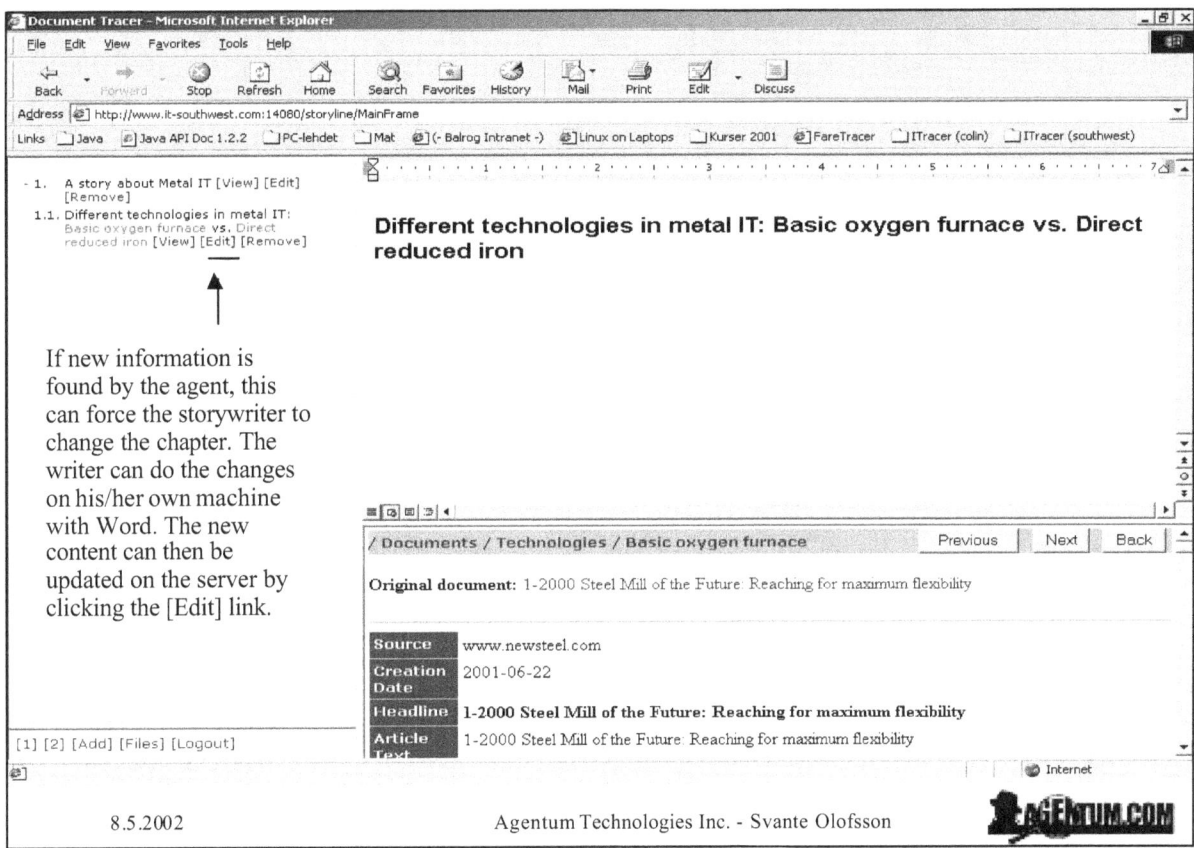

If new information is found by the agent, this can force the storywriter to change the chapter. The writer can do the changes on his/her own machine with Word. The new content can then be updated on the server by clicking the [Edit] link.

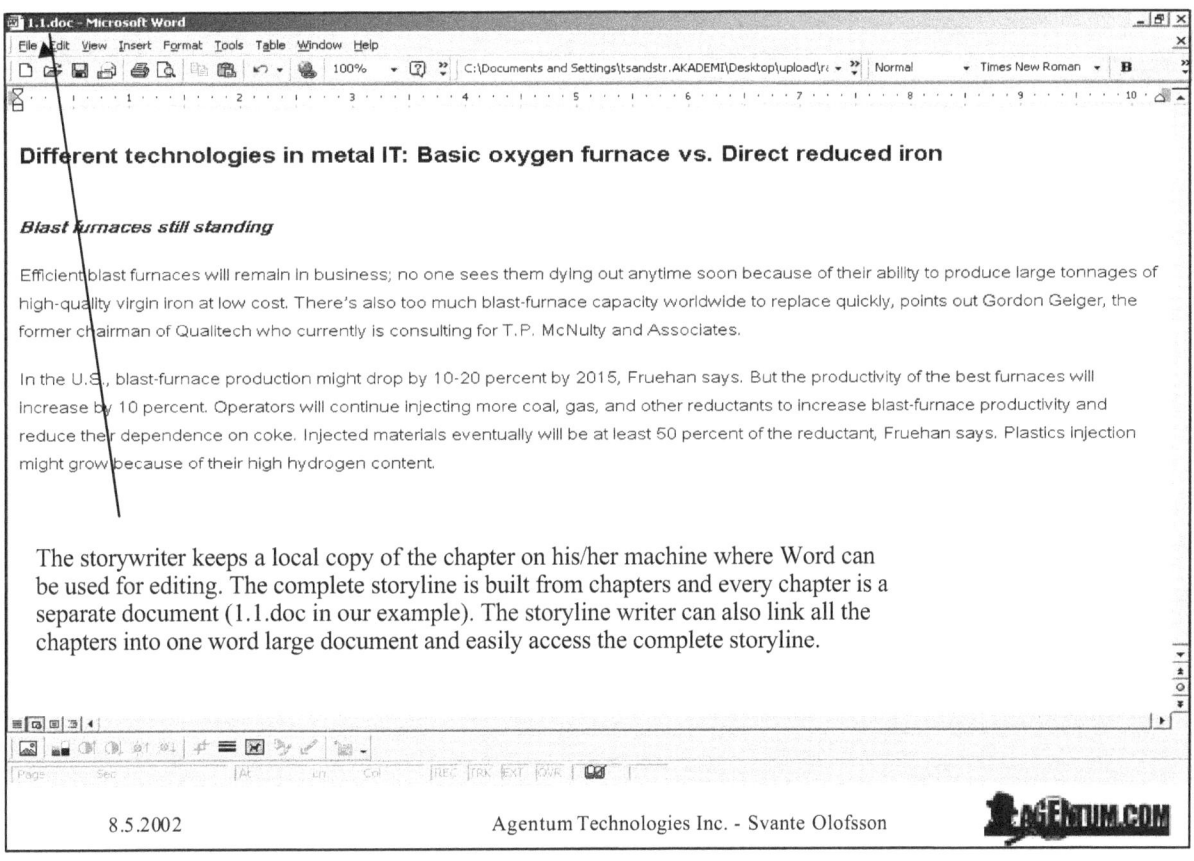

The storywriter keeps a local copy of the chapter on his/her machine where Word can be used for editing. The complete storyline is built from chapters and every chapter is a separate document (1.1.doc in our example). The storyline writer can also link all the chapters into one word large document and easily access the complete storyline.

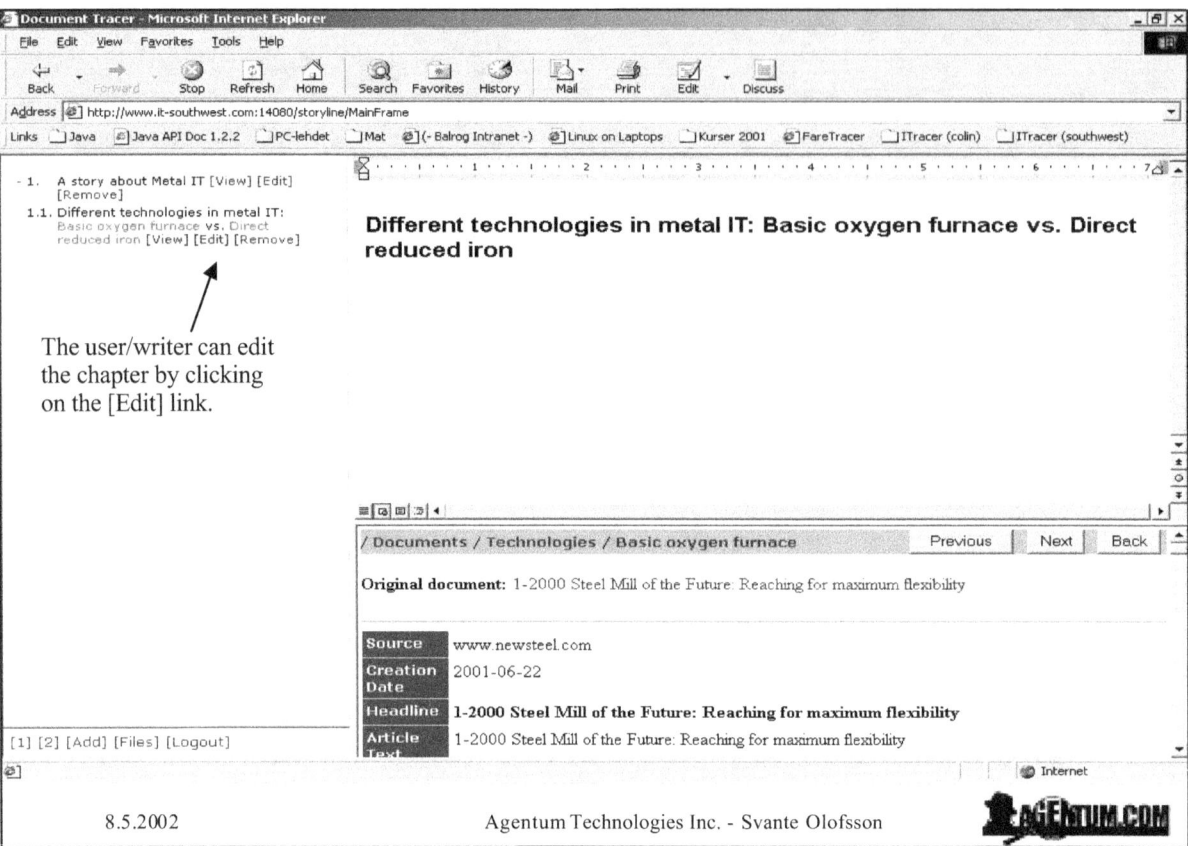

The user/writer can edit the chapter by clicking on the [Edit] link.

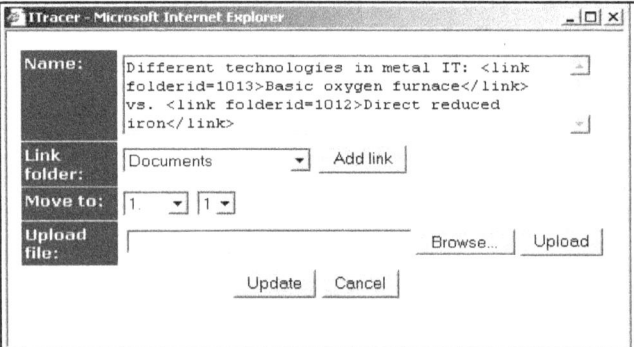

1. The name can be changed by editing the *Name:* field.
2. The linking of folders is done by selecting the folder to link to and then clicking on the [Add link] button. After the button is pressed a '<link folderid=xxxx>*folder name*</link>' entry is added to the name field.
3. The headline can also be moved to another chapter if the storywriter notices that the chapter is better suited in another part of the storyline.
4. The user can select the edited document (1.1.doc) by pressing the [Browse] button and then upload it to the server by pressing the [Upload] button.

When the storywriter is satisfied with the changes he/she can store them by pressing the [Update] button.

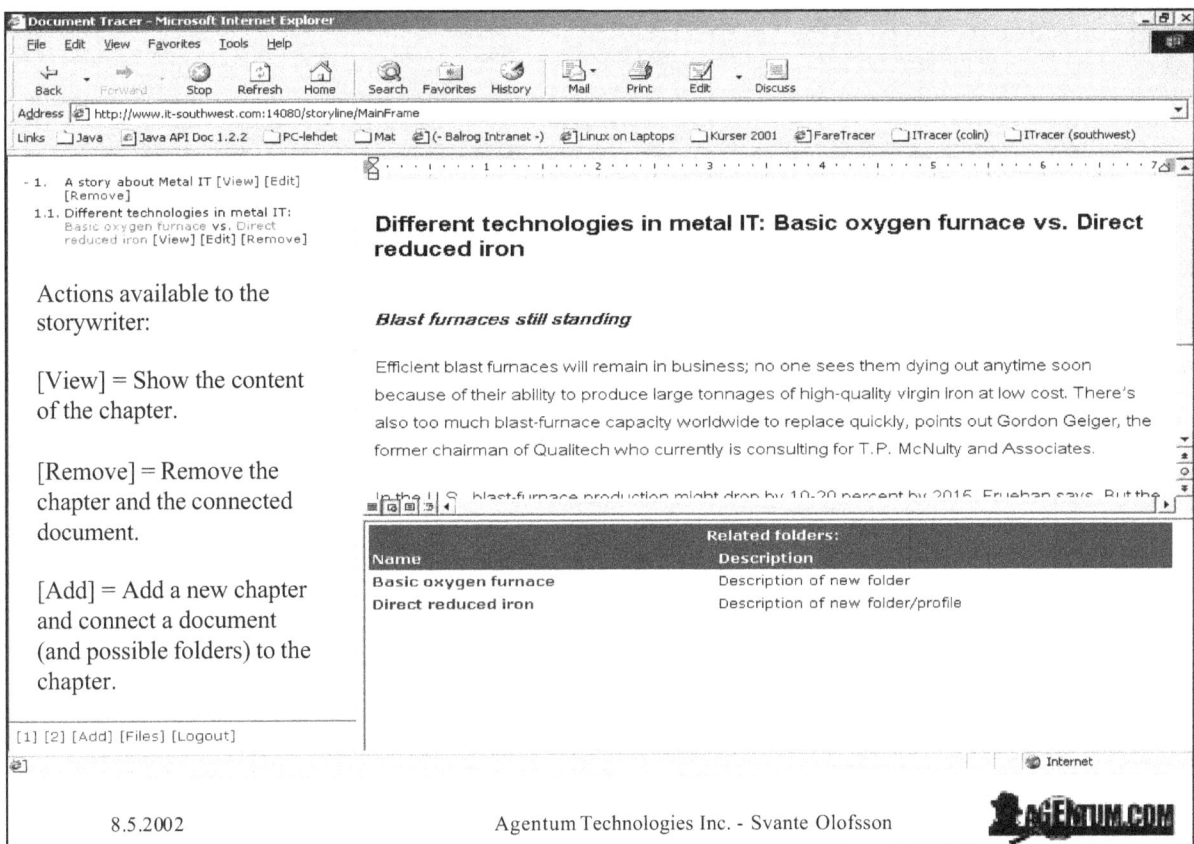

Actions available to the
storywriter:

[View] = Show the content
of the chapter.

[Remove] = Remove the
chapter and the connected
document.

[Add] = Add a new chapter
and connect a document
(and possible folders) to the
chapter.

8.5.2002   Agentum Technologies Inc. - Svante Olofsson

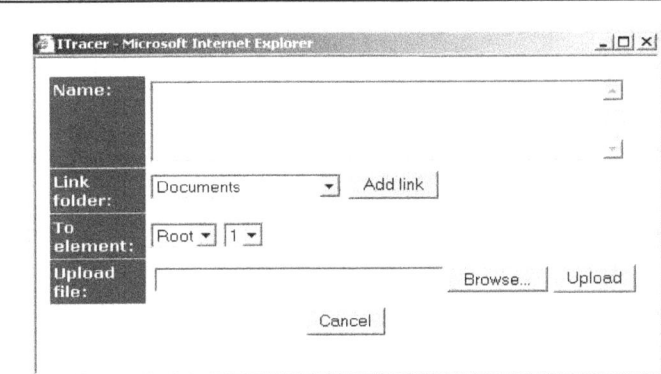

Adding of a chapter opens up the same dialog box as the [Edit] link did. The name of the
new chapter is entered into the name field together with links to agent folders.

Next step is to choose the place for the new chapter. The chapter is linked to the correct place in
the storyline by selecting the element the new chapter should belong to. The number of the chapter
is set by selecting the order (one is the first and two is the second, etc...) of the new chapter under
the element.

The last step is to add the document that contains the chapter content. This is done by selecting the
file to upload with the [Browse] button and then sending it to the server with the [Upload] button.

When the above is done a [Add] button will appear and the chapter added when it is pressed.

8.5.2002   Agentum Technologies Inc. - Svante Olofsson   

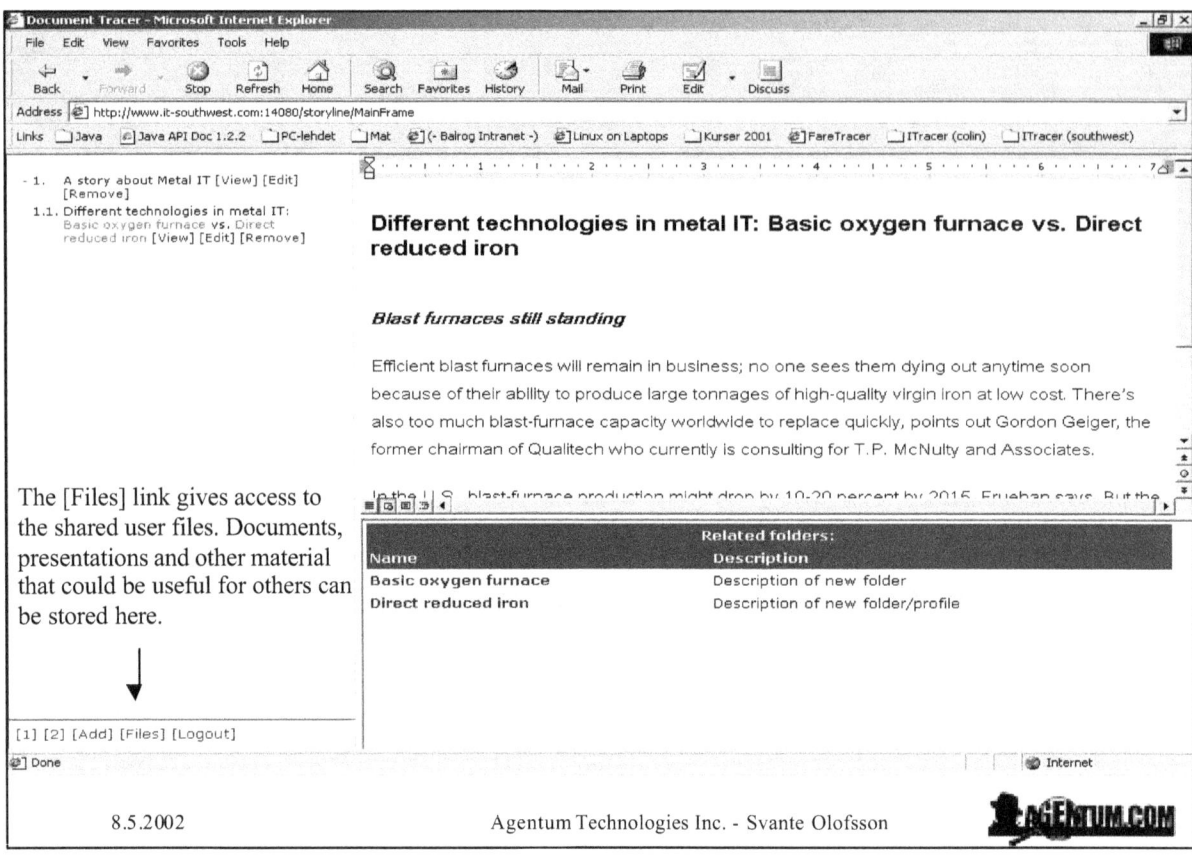

The [Files] link gives access to the shared user files. Documents, presentations and other material that could be useful for others can be stored here.

8.5.2002   Agentum Technologies Inc. - Svante Olofsson

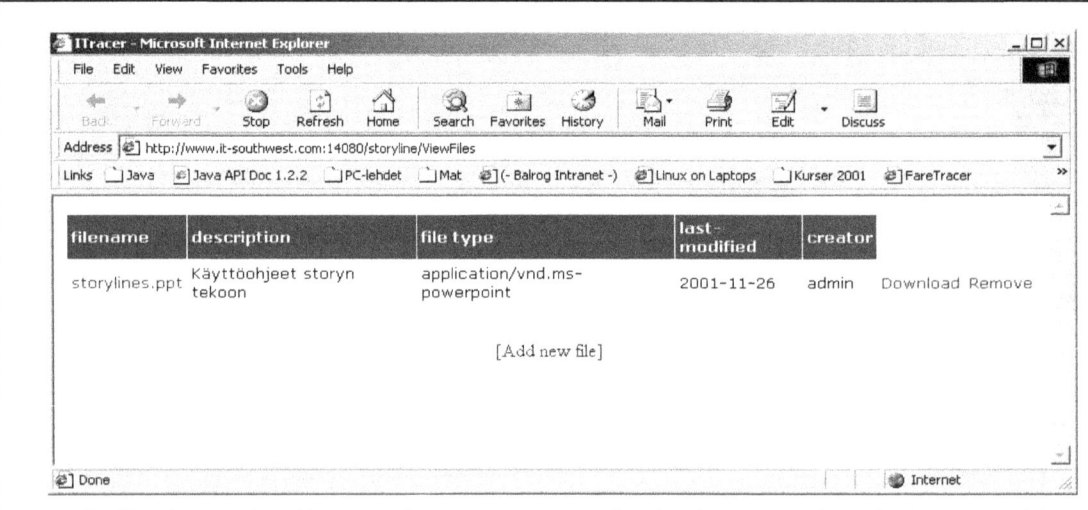

The files that are shared between the users are presented as the above example. Only the owner of the file can remove it. The files can be opened directly in Explorer by clicking on the filenameit. There is also a [Download] link that should be used for large files and slow connections (modems). The [Download] link compresses the file as a zip file before it is downloaded.

A user can add a file by clicking on the [Add new file] link. A short description can be added so that the other users get a better idea what the file contains.

8.5.2002   Agentum Technologies Inc. - Svante Olofsson

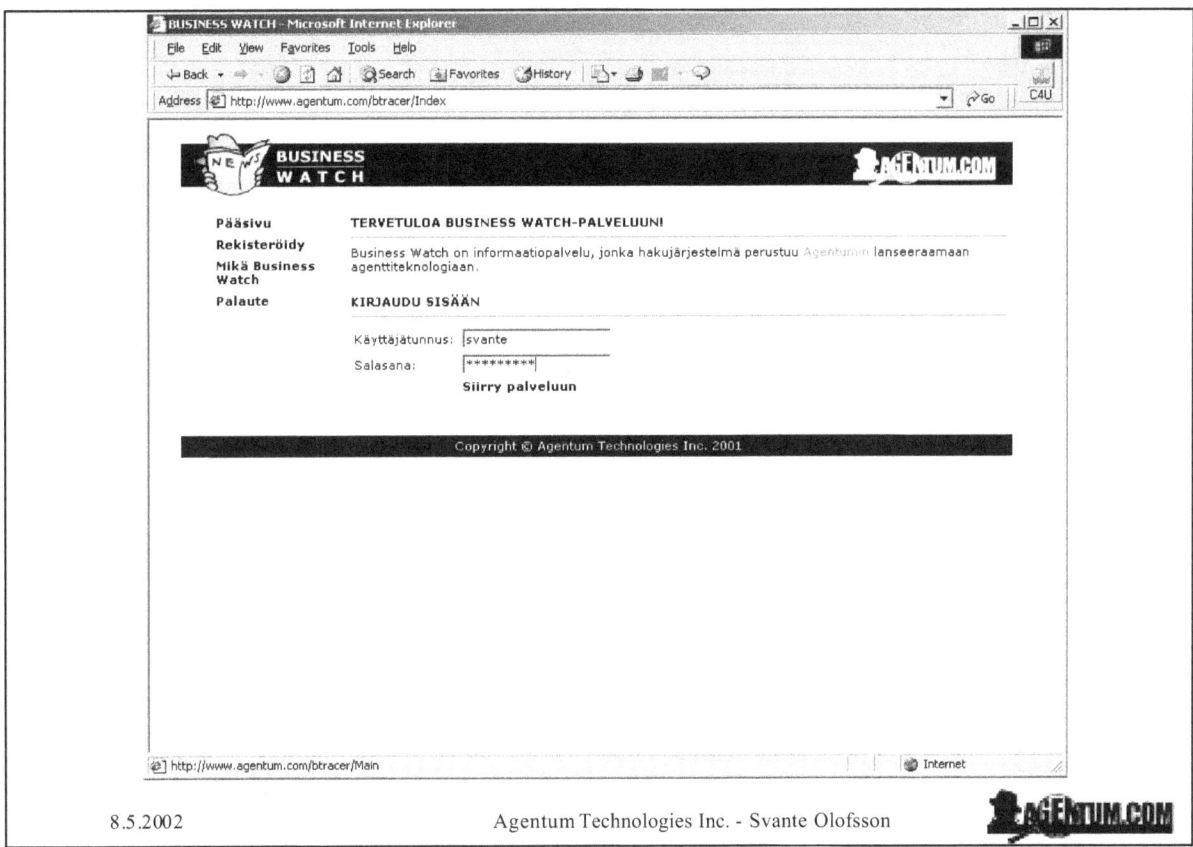

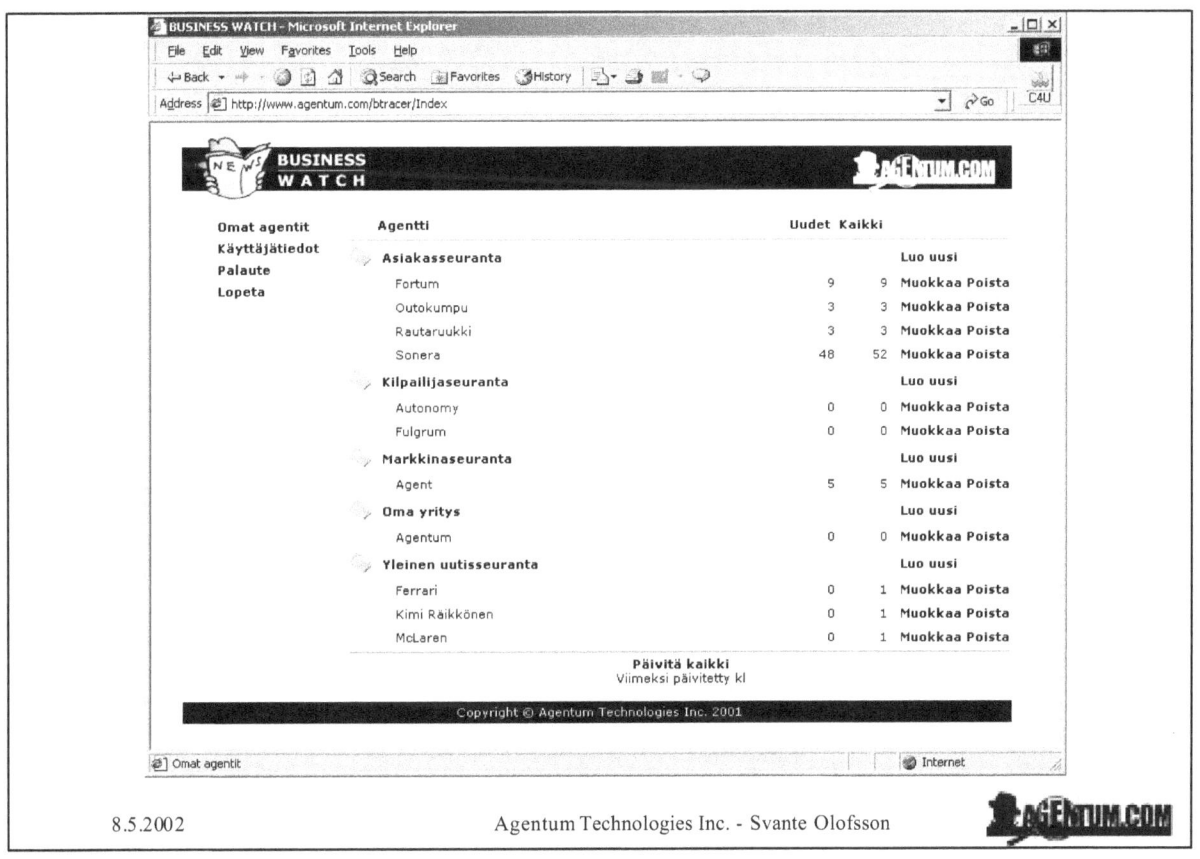

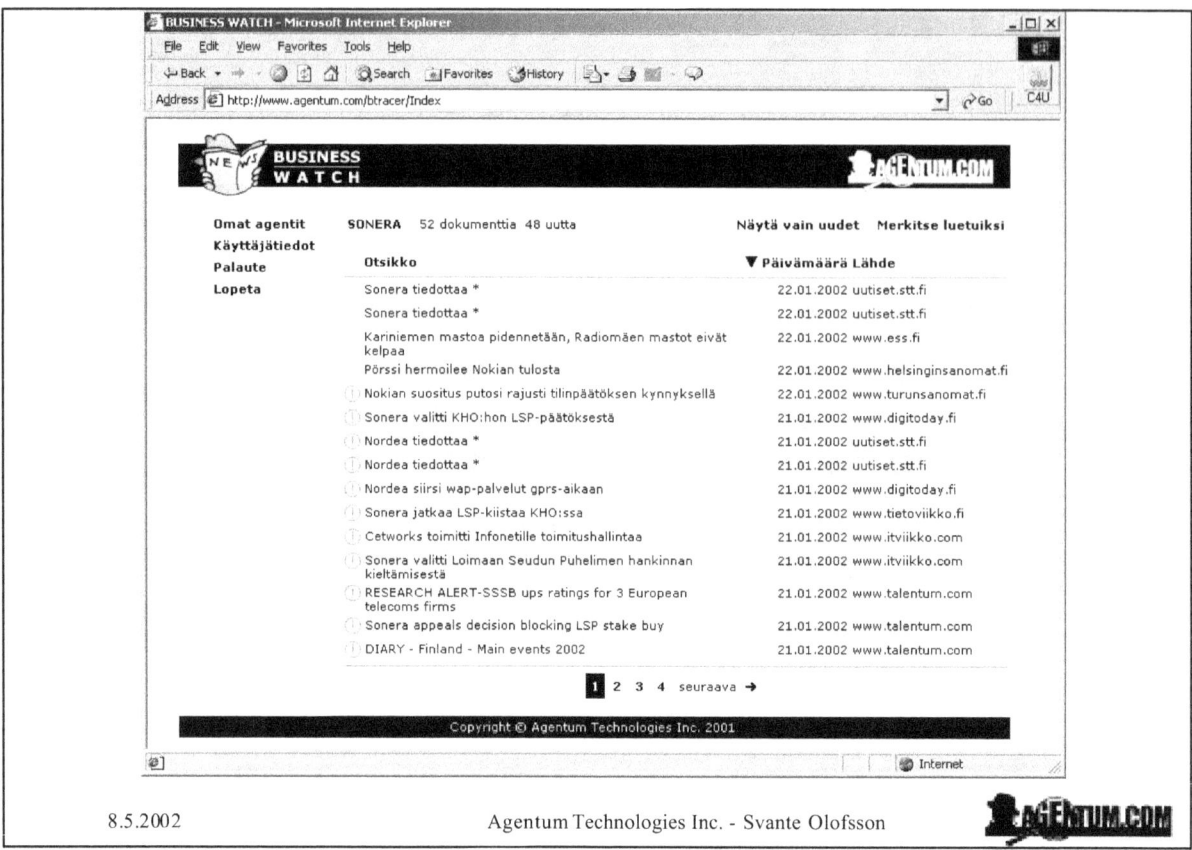

8.5.2002   Agentum Technologies Inc. - Svante Olofsson

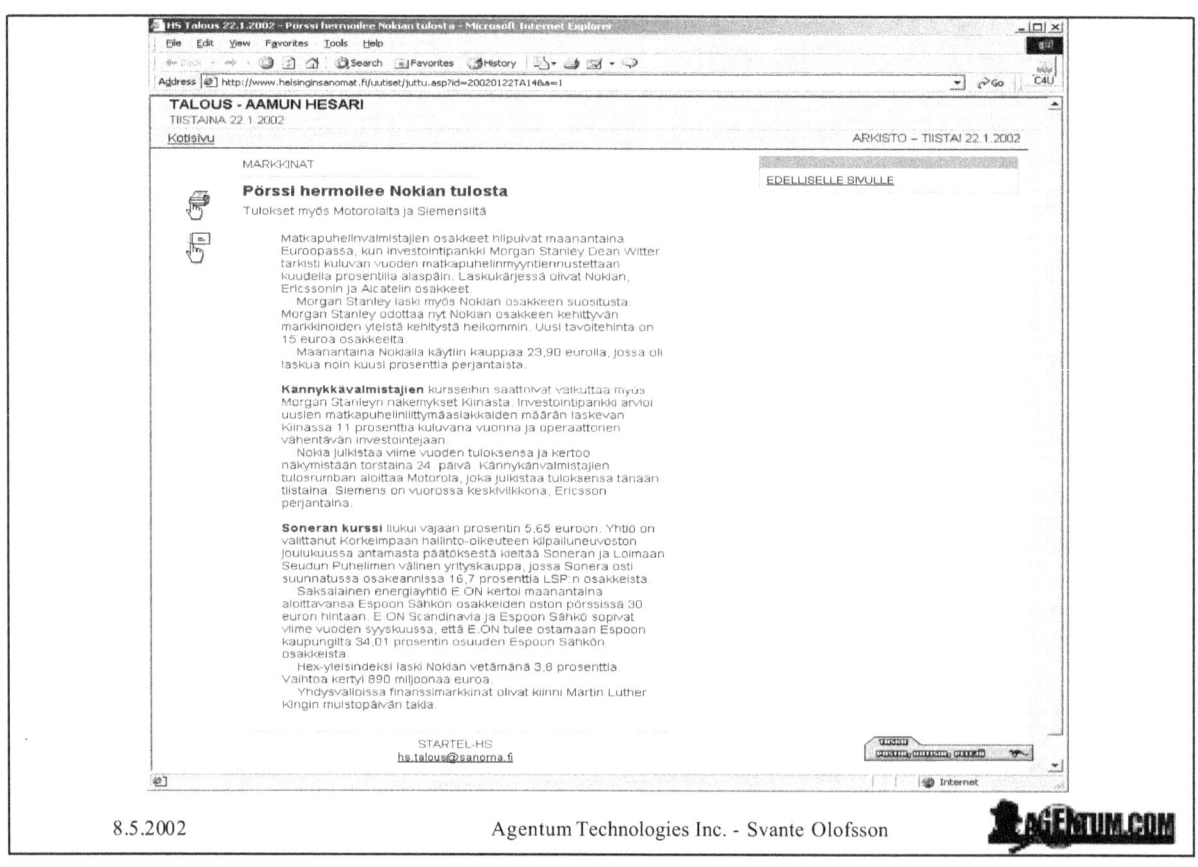

8.5.2002   Agentum Technologies Inc. - Svante Olofsson

DAY 3

SESSION 3

Mikael Collan
"A Method for Including Foresight Information into Fuzzy Real Option Valuation"

Shuhua Liu
"Intelligent Agents"

A METHOD FOR INCLUDING DYNAMIC TREND INFORMATION IN FUZZY PRICING OF REAL OPTIONS

Mikael Collan
Åbo Akademi University
IAMSR / TUCS
Turku Center for Computer Science

Peter Majlender
Åbo Akademi University
IAMSR / TUCS
Turku Center for Computer Science

ABSTRACT

This paper presents a heuristic method for dynamically including external (trend) information to pricing real options with a fuzzy presentation of the Black and Scholes option pricing formula. The method manipulates the tails of trapezoidal fuzzy cash flow estimates to incorporate trend information into capital budgeting with fuzzy cash flows.

Keywords: option pricing, fuzzy sets

JEL classification: C0, G3

Turku Centre for Computer Science
Lemminkäisenkatu 14A
20520 Turku
Finland

IAMSR
Åbo Akademi University
Lemminkäisenkatu 14B 6th floor
20520 Turku
Finland

INTRODUCTION

Capital budgeting for investments with a long economic life is difficult for a number of reasons. The further into the future the cash flows expand the more difficult it is to assess them accurately. Information about the future is harder to obtain, and more difficult to integrate to plans, because information from far away in the future will most probably be in qualitative rather then quantitative form. When the economic life of an investment is long, also managerial performance becomes an important issue, managerial flexibility to steer the investment will certainly have a value.

Flexibility in capital budgeting is an issue that has been talked about a lot in the last few years, and real options have established their position as a way to assess the value of managerial flexibility. The valuation of real options is based on the option pricing models originally created for pricing of financial options. Commonly two methods are used, the binomial option pricing formula and the Black and Scholes option pricing formula created by Black and Scholes (1973), and further modified by Merton (1973). Real options is a term first introduced in Myers (1977) article. The focus of research on real options has been shifting from general presentations about managerial flexibility, Trigeorgis and Mason (1987), Kulatilaka and Marks (1988), and Sick (1989), and general theory of real options, Trigeorgis (1988), Triantis and Hodder (1990), and Dixit and Pindyck (1994) to exploring advanced issues in the valuation of real options, Trigeorgis ed. (1995) and Alvarez and Stenbacka (2000), and consideration about practical issues of real option pricing, Triantis (1999), Amram and Kulatilaka (1999), and Copeland and Antikarov (2001).

Because valuation of real options is valuing contingent uncertain outcomes in the future, issues that have to do with information about the future should also be included in the valuation process. Obtaining managerial information by looking at the environment is commonly called environmental scanning. A book by Aguilar (1967) presents the theoretical basis for the information needs of a manager. Modern information technology enables the environmental scanning to be made effectively because of the information explosion (high availability of relevant information) and sophisticated tools to extract information, Zeleny ed. (2000) and Liu (2000). As already discussed earlier, the information about future events is often in qualitative form, thus it is important to be able to use the qualitative information in the capital budgeting process. Integrating qualitative information to financial calculations has been discussed, for example, in Hamscher, Kiang, and Lang (1995), Alpar and Dilger (1995), Benaroch and Dhar (1995), and Feelders and Daniels (2001).

Because expected values of future cash flows are difficult to assess, especially in connection with investments with a long economic life, presenting them with a single number and with probability theory is in many cases unrealistic. A better, and more sophisticated approach is to estimate the expected cash flows by fuzzy numbers, and with possibility theory. Fuzzy sets, introduced by Zadeh (1965) and possibility theory have been applied to capital budgeting by Buckley (1987), Li Calzi (1990), Carlsson and Fullér (1999), Kuchta (2000), and Kahraman, Ruan and Tolga (forthcoming). Inclusion of qualitative information into bank operations with fuzzy constraints has been presented in Gardin, Power, and Martinelli (1995).

Fuzzy presentation of option pricing in connection with the binomial option valuation formula is presented in Muzzioli and Torricelli (2000), and in Muzzioli and Torricelli (2001). Fuzzy real option valuation has been presented in Carlsson and Fullér (2000A) and in Carlsson and Fullér (2000B). Carlsson, Fullér, and Majlender (2001) have examined capital project selection with fuzzy real options.

The contribution of this paper is to present a method of integrating qualitative future information into capital budgeting that can be used in a fuzzy version of Black and Scholes real option model. The method incorporates fuzzified estimates of expected revenues and costs, with the possibility to include external (e.g. trend) information dynamically into the estimates. The inclusion of external information is achieved by introducing a heuristic method to adjust the fuzzy representations of the cash flow estimates (the trapezoidal fuzzy numbers) according to the information.

The composition of this paper is the following. Firstly, the fields of study that the method is based on are shortly introduced. Secondly, the method is presented. Thirdly, the presented method is discussed and propositions for future studies on the subject shortly elaborated. Finally, the ideas presented are summed up.

THE METHOD

The method is devised to facilitate inclusion of (external) trend information into a series of fuzzy cash flow estimates used in capital budgeting. The idea behind the method is that trends can be modelled by changing the shape of the trapezoidal fuzzy numbers that represent the cash flow estimates. For a graphical definition of the trapezoidal fuzzy numbers see figure 1 below.

Figure 1. definition of the fuzzy cash flow estimates

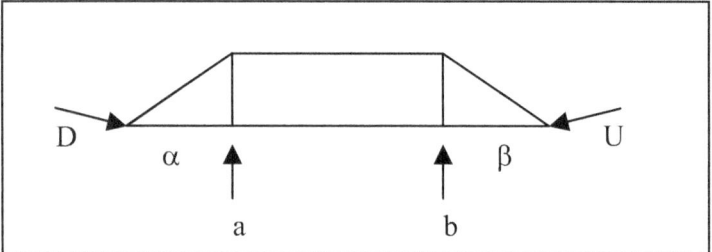

To be able to manipulate the trapezoidal fuzzy numbers under all circumstances we use an unorthodox heuristic method and simply consider the tails of the fuzzy sets. To manipulate the tails we simply detach them from the core, then manipulate them according to the trend information, and finally re-attach them to the core. The tails of the trapezoidal fuzzy numbers are called α and β and their size (length) is defined by:

Size (length) of $\alpha = a - D$

Size (length) of $\beta = U - b$

In order to consistently model trends we need to have rules according to which the tails are adjusted. We need to know the size of the tails and how the size should be distributed between them.

The method is based on the assumption that when a trend is positive it can be modelled by adjusting the right tail (β) of a trapezoidal fuzzy number towards the right (positive) to capture the positive change. At the same time, however, the left tail (α) will be adjusted as well, because we assume that if the possibility of a positive outcome grows, then at the same time the possibility of a negative outcome declines. The inverse applies for a negative trend. If the trend is neutral (or the trend is not known) then the tails are of an equal size.

To determine the relation of the size of the tails our method uses a system of first dividing the total length of the tails ($\alpha + \beta$) into n equally large units and then distributing all the units between the tails according to the effect of the trend. This means that the number of units represents a number of degrees the effect of a trend can have. The larger the number of units the more accurate the modelling of the trends can be.

The above means that:

$\alpha < \beta$ for a positive trend

$\alpha > \beta$ for a negative trend

$\alpha = \beta$ for static trend or for when trend not known

and that:

$\alpha + \beta = n$, where n $\geq$ 0

To define the size of the tails we use the average of the extreme values of the core (average of a and b) as a proxy for the size of the tails. We have selected the average of a and b, because we feel that it links the size of the tails to the size of the core in a good proportionate way. However, any other value from between a and b can be used as well. An exogenous value is also possible, for example, a value derived from the volatility of the stream of cash flow estimates. We apply the relation between the sizes of the tails as a percentage of the number of units allocated to each tail of the total number of units (n):

To define the maximum value for the downside (D) we use:

$$D = a - \frac{n_\alpha}{n} * \frac{a+b}{2}$$

where n_α is the number of units allocated to α-tail

To define the maximum value for the upside (U) we use:

$$U = b + \frac{n_\beta}{n} * \frac{a+b}{2}$$

where n_β is the number of units allocated to β-tail

With the operations described above we can define the tails of a trapezoidal fuzzy number to represent the effect of trends. By re-attaching the tails to the core we have a trend adjusted trapezoidal fuzzy cash flow estimate. As trend information changes the operation may be performed again to reflect the latest information, making the method allow dynamic trend adjusted fuzzy pricing of real options.

DISCUSSION

Investments that have a long economic life are sensitive to changes in the market factors, therefore, it makes sense to include the effects of the changes in the calculations as soon as they are known. However, making changes is not a very easy task in a situation where the planning and market-analysis are separated. This is to say, that the people who plan investments are not the same people who will have the information about the changes of trends in the future. Also, if the economic life of an investment is very long, then the original planners are most probably no longer available to explain the original plans. This also means that the people who have the information about the changes in trends are people, who may not know the technical background of the plans, and, therefore, the adjusting of cash flow estimates is made more difficult. We dare to suppose that the core of the fuzzy cash flow estimates includes most of the technical variability of the project, which would suggest that it is rather safe for market analysts to adjust the tails of the estimates. The point made above is not trivial, because it overcomes some of the problems with separation of operational management and strategic planning of investments.

Information about the future often comes in the form of qualitative rather than in a quantitative form, qualitative information should be included if possible. As the method includes a system of degrees to model the effect of the trends it makes possible the inclusion of qualitative information. The effect of a trend needs to be assessed by an expert, for example, a market-analyst, who will assign the degree of effect of the trend. As intelligent agents evolve it will be possible to automate the process of adjusting the fuzzy cash flow estimates automatically. This is an interesting prospective and certainly worth further study.

SUMMARY

This paper has presented a method to include trend information to capital budgeting by way of adjusting the tails of trapezoidal fuzzy cash flow estimates. The cash flows can further be used in fuzzy real option valuation.

The method escapes some problems of temporal and physical separation of planning and market analysis and offers a simple and straightforward way to including both qualitative and quantitative data into capital budgeting. The method suits large investments with a long economic life.

The paper suggests the following type of approach to include trend information into capital budgeting:

1. Analysis of market information to find trends
2. Assigning the trend a degree according to the selected degree system
3. Adjustment of the trapezoidal fuzzy number according to the method
4. Using the adjusted fuzzy numbers in capital budgeting (real option pricing)

A software platform using the method presented in this paper is under construction by the authors.

REFERENCES

Aguilar, F.J., 1967, Scanning the Business Environment, McMillan

Alpar, P. and Dilger, W., 1995, Market share analysis and prognosis using qualitative reasoning, Decision Support Systems, 15(2), pp. 133-146

Alvarez, L.H.R. and Stenbacka, R., 2000, Adoption of uncertain multi-stage technology projects: a real options approach, Journal of Mathematical Economics

Amram, M., and Kulatilaka, N., 1999, Real options: Managing strategic investments in an uncertain world, Harvard Business School Press, Boston, Massachusetts

Benaroch, M. and Dhar, V., 1995, Controlling the complexity of investment decisions using qualitative reasoning techniques, Decision Support Systems, 15(2), pp. 115-131

Black, F. and Scholes. M., 1973, The pricing of options and corporate liabilities, Journal of Political Economy, 81, 637-659

Buckley, J.J., 1987, The fuzzy mathematics of finance, Fuzzy Sets and Systems, 21, pp.257-273

Carlsson, C. and Fullér, R., 1999, Capital budgeting problems with fuzzy cash flows, Mathware and Soft Computing, 6, pp.81-89

Carlsson, C. and Fullér, R., 2000A, On fuzzy real option valuation, TUCS Technical Report No 367, October, Turku Centre for Computer Science

Carlsson, C. and Fullér, R., 2000B, Real option evaluation in fuzzy environment, Proceedings of the International Symposium of Hungarian Researchers on Computational Intelligence, Budapest Polytechnic, pp. 69-77

Carlsson, C., Fullér, R., and Majlender,P., 2001, A possibilistic approach to selecting portfolios with highest utility score, Fuzzy Sets and Systems

Copeland, T. and Antikarov, V., 2001, Real options: A practitioner's guide, Texere

Dixit, A.K. and Pindyck, R.S., 1994, investment under uncertainty, Preston, NJ, Princeton University Press

Feelders, A.J., and Daniels, H.A.M., 2001, European Journal of Operational Research, 130(3), May, pp. 623-637

Gardin, F., Power, R., and Martinelli, E., 1995, Liquidity management with fuzzy qualitative constraints, Decision Support Systems, 15(2), pp. 147-156

Hamscher, W., Kiang, M.Y., and Lang, R., 1995, Qualitative reasoning in business, finance, and economics: Introduction, Decision Support Systems, 15(2), pp. 99-103

Kahraman, C., Ruan, D., and Tolga, T., forthcoming, Capital budgeting techniques using discounted fuzzy versus probabilistic cash flows, Information Sciences

Kuchta, D., 2000, Fuzzy capital budgeting, Fuzzy Sets and Systems, 111, pp. 367-385

Kulatilaka, N. and Marks, S., 1988, The strategic value of flexibility: Reducing the ability to compromise, American Economic Review, pp. 574-80

Li Calzi, M., 1990, Towards, a general setting for the fuzzy mathematics of finance, Fuzzy Sets and Systems, 35, pp. 265-280

Liu, S., 2000, Improving executive support in strategic scanning with software agent systems, Ph.D. dissertation, Åbo Akademi University

Merton, R., 1973, Theory of rational option pricing, Bell Journal of Economics and Management Science, 4, 141-183

Muzzioli, S. and Torricelli, C., 2000, Combining the theory of evidence with fuzzy sets for binomial option pricing, Materiale di discussione n. 312, Dipartimento di Economia Politica, Universitá degli Studi di Modena e Reggio Emilia, May

Muzzioli, S. and Torricelli, C., 2001, A model for pricing an option with a fuzzy payoff, Fuzzy Economic Review, vol. VI, number.1, May.

Myers, S.C., 1977, Determinants of corporate borrowing, Journal of Financial Economics, 5(2), pp. 147-175

Sick, G., 1989, Capital budgeting with real options, Salomon Brothers Center, New York University

Triantis, A.J., 1999, Real options in corporate risk management, in Corporate Risk: Strategies and Management, ed. Brown, G.W. and Chew, D.H., Risk Books, London

Triantis, A.J. and Hodder, J., 1990, Valuing fexibility as a complex option, The Journal of Finance, XLV(2), June, pp. 549-65

Trigeorgis, L., 1988, A conceptual options framework for capital budgeting, Advances in Futures and Options Research, 3, pp. 145-67

Trigeorgis, L. ed., 1995, Real options in capital investment: Models, strategies, and applications, Westport, Connecticut, Praeger

Trigeorgis, L. and Mason, S.P., 1987, Valuing managerial flexibility, Midland Corporate Finance Journal, 8(1), Spring, pp. 14-21

Zadeh, L.A., 1965, Fuzzy Sets, Information and Control, 8, 338-353.

Zeleny, M. ed., 2000, The IEBM Handbook of Information Technology in Business, Business Press, Thomson Learning

Shuhua Liu
Institute for Advanced Management Systems Research
ÅBO AKADEMI UNIVERSITY

Intelligent Agents Supported Real Option Valuation

ROW – International Real Option Workshop
Turku, Finland, 6-8.5. 2002

Contents

- Investment Decision Making with the Real Option Approach
- Real Option Valuation: A Process View
- Need for Support Technology
- Intelligent Agents
- Agents Supported Real Option Valuation

Large Investments

- Strategic importance
- Long life cycle
- Risks and potentials unknown at the beginning
- The nature of the venture may change during its life.
  - end product market, even the technology base, can probably change several times

Real Option Valuation for Investment Decision Making (1)

- Real Option Valuation Approach:
  - Throughout a project's life cycle, managers can make a series of "scale-up or not" evaluation and decisions similar to the multiphase reviews at most venture capital firms (Dahlberg and Porter, 2000).
  - By phasing the projects, every step in a project opens or closes the possibility for further options. Creating options can buy time to think and gain information to decide for the next move (Collan, Carlsson and Majlender, 2002) .

Real Option Valuation for Investment Decision Making (2)

- Real Option Valuation Approach:
  - On-going learning about the risks and potentials of a new venture over time; adaption of actions.
  - Considers future uncertainty and express it in terms of alternative options that are evaluated by the anticipated benefits or potential payoffs rather than the risks. (Dahlberg and Porter, 2000).

Real Option Valuation Approach: A Process View

- Real Option Valuation is a repeated process of:
  - Identifying options (the multiple paths to success) in light of newly accessed information: develop the decision trees at different project stage
  - Evaluating the options: quantitative and qualitative analysis of the value of the options (computing the decision tree)
  - Selecting the options: ranking of or voting for real options based on the valuation (find the critical / optimal path)
  - Executing the immediate option and start the next round ROV process

The Need for Decision Support Technology

- Investment management as a project review process in which ongoing evaluation effort is structured around key decision points, or triggered by changes in the business environment.

- The valuation process, even the computing process, are not intended to provide a definite answer but rather to provide decision makers an on going dialogue about the project. (Dahlberg and Porter, 2000).

- This requires several things:

The Need for Decision Support Technology

- up-to-date project status information readily available to decision makers;

- up-to-date market information or industry foresights (current events or trends) be made aware to decision makers constantly and be integrated into the various phases of a ROV process;

- option analysis and evaluation be done periodically or event-triggered, applying advanced option valuation methods;

- analytical results be explained using easily understandable business terms

Intelligent Agents

- Central to the notion of software agents are the automation of work and the automation of computer usage.

- Software agents are computational programs that:
  - inhabit in a computing environment
  - act on behalf of users to accomplish delegated tasks
  - decide own course of actions dynamically while responding to the environment (Maes, 1994; Wooldridge and Jennings, 1995)

Characteristics of Software Agents

- Situatedness: receives sensory input from its environment and perform actions that change the environment in certain ways (reactive and responsive)
- autonomous: able to take initiative, to solve problem without direct intervention and constant guidance from the user
- adaptive and intelligent: to customize its assistance and service according to what it learns about the user; to improve its performance based on previous experience
- proactive support and service
- work in background, serve round-clock

Software Agents vs Traditional Computing Programs

- Computing programs traditionally depend on users to use them. They usually remain dormant until specifically called by user instructions.

- Software agents do not rely on users' explicit action to be activated and directed step by step. Agents make it possible for the programs to work independent of users' presence and instructions.

- Agents can deliver customized, user-wanted information and service to the user.

- Agents can interact with user and with each other.

Benefits from Agent Applications

- An alternative interfacing approach
  - Ask-and-Delegate vs Direct Manipulation
  - saves decision makers time and unnecessary system operating efforts.
  - Task and responsibility delegation support user mobility;
- Encourages user empowerment, bypass intermediaries between the application and end user, thus eliminate delay in the process and free up human resources.
- Proactive information delivery.
- Agent approach offers an alternative abstraction means used to manage complexity, and to conceptualize, design, implement complex systems or IS applications.
- Flexible system to meet changing requirements

Intelligent Agent Supported
Real Option Valuation: A Framework

- Support components for ROV process:
    - Scanning Agent: support the data collection process
    - Interpretation Agent: support data pre-analysis
    - Option Generator: help to identify and generate options
    - Option Valuator and Selector: help to calculate option values using selected valuation methods
    - Project Reviewer: document the project history and current status: events, milestones, accumulated knowledge, etc.

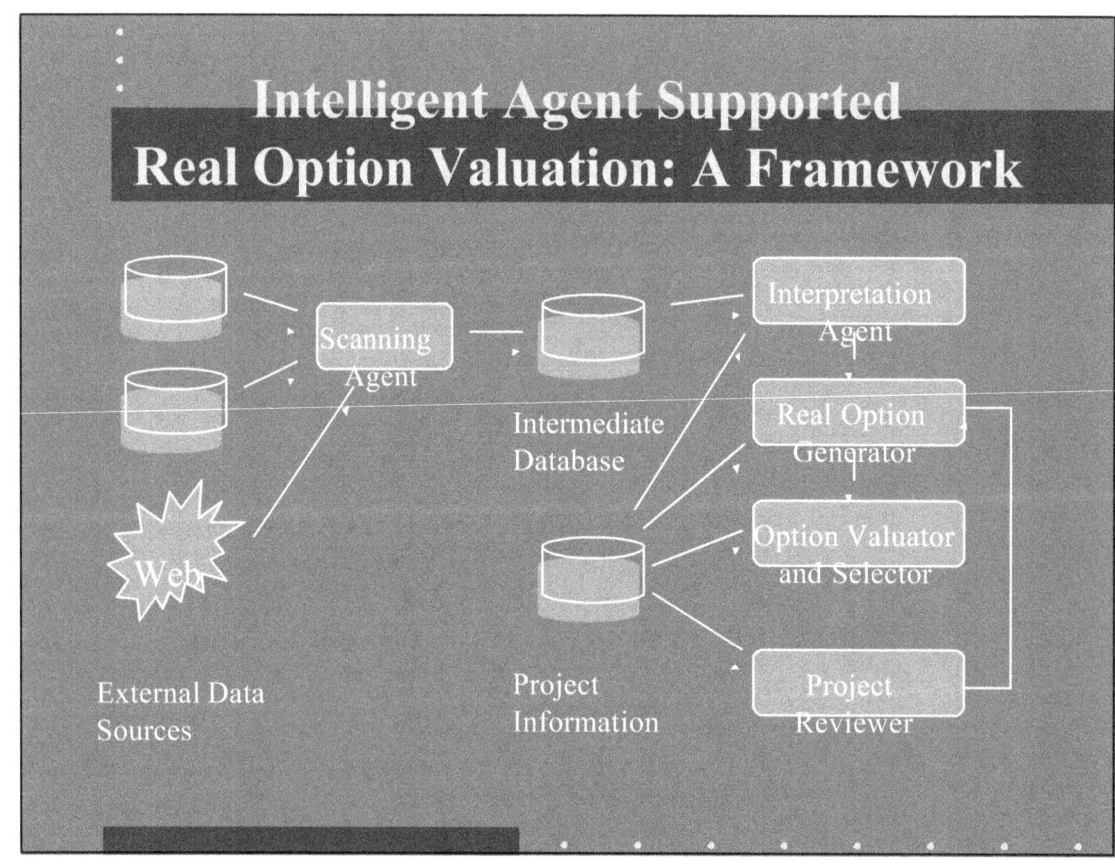

Intelligent Agent Supported
Real Option Valuation: A Framework

Human Scanning Activities

- Tend to be intuitive and fragmented; difficult to be systematic;
- Continuous scanning activities are very time consuming and cost a lot;
- However there is a need for information to be updated constantly and in time.
- The speed of access to information is also critical.

Scanning Agent

- Data collection: retrieval with filtering, informing
- Agent Server
  - Watch selected data sources for new information and retrieves only relevant information, which is specified by a user profile defined by a set of factors.
  - Transform the collected information from different data sources into a consistent format and store the information in a intermediate data storage.
  - Allows commenting (on articles and reports) for later reviewing;
  - Supports adding/removing of sources;
- Agent Client: interface, agent control and configuration

Scanning Agent: Benefits

- Increase current awareness of what is happening in the business environment, by advertising incoming information and data are updated automatically as often as necessary;
- Continuous and systematic scanning;
- Frequent updating
- Avoid manually-made mistakes;
- Manpower Saving.

Interpretation Agent

- The need for timely analysis and interpretation of the collected data
- Agent Server
  - Identify business events and trends
  - Draw structured information from text resources

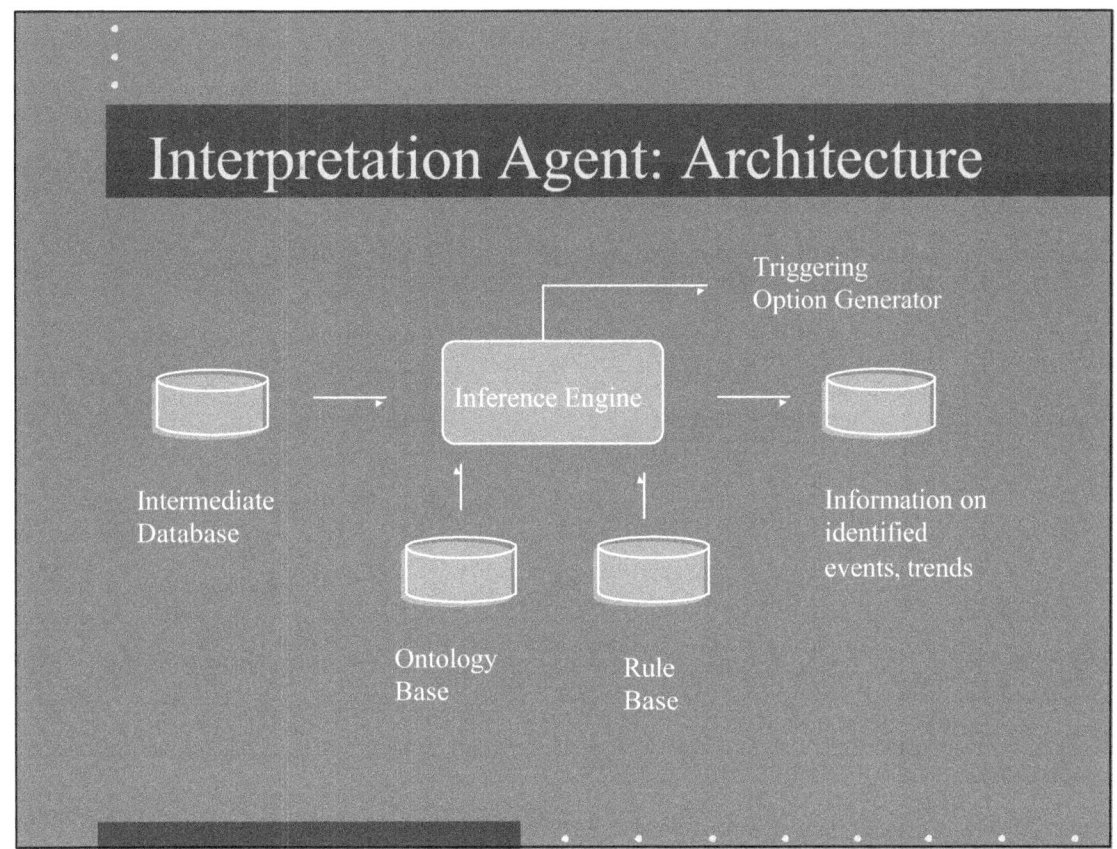

Interpretation Agent: Architecture

Intermediate Database → Inference Engine → Triggering Option Generator

Ontology Base, Rule Base, Information on identified events, trends

Option Generator

- Triggered by the Interpretation Agent or Project Reviewer;
- Supported by pre-developed various types of possible options stored in the Project Information base;
- To generate a draft decision tree for a decision maker to confirm or modify;
- Make the decision tree available to Option Evaluator and Selctor

Option Valuator and Selector

- Triggered by Option Generator or the user;
- Supported directly by project history and status information;
- Supported by pre-developed valuation methods embedded in it;
- Compute the option values.
- Rank and select among alternative options based on user specified criteria.
- Add the new analytical results into project information base and communicate the analytical result in directly useful and understandable terms.

Project Reviewer

- Project Information Base
  - Types of possible real options
    - option to invest
    - option to proceed
    - option to discontinue
    - option to wait
    - option to split
    - option to give up
    - option to abandon
    - option to postpone
    - option to modify
  - Option objects
    - name, value, time to maturity
    - value lost over duration of the option
    - deciding factors
    - description

Project Reviewer

- Sits in the Project Information base;
- Be awared of the last specified project milestones and check the project status periodically or on request.
- Alert on delays, problems, newly available options, and so on.

Organised by

Institute for Advanced Management Research
Åbo Akademi University
Lemminkäinengatan 14 B
FIN-20520 Åbo, Finland

Celebrating

75th Anniversary of Business Education at Åbo Akademi University

In cooperation with

Turku Centre for Computer Science

European Network of Excellence on Intelligent Technologies for Smart
Adaptive Systems

eunite

www.ingramcontent.com/pod-product-compliance
Lightning Source LLC
Chambersburg PA
CBHW080636180526
45168CB00008B/3184